Engaging Learners

in Building Cognitive Functions that Achieve Mathematical Academic Standards

James T. Kinard, PhD

Gwendolyn D. Gibson Kinard, PhD

RMT Laboratory Inc.
P.O. Box 803934
Chicago, Illinois 60680
www.rmtlabinc.com

ISBN-13: 978-0692078280
ISBN-10: 0692078282

Printed in the United States of America by Create Space, an Amazon.com Company

Advance Review and Early Praise for
Engaging Learners in Building Cognitive Functions that Achieve Mathematical Academic Standards

Dr. James Kinard and Dr. Gwendolyn D. Gibson Kinard have written a remarkable book relating to adopting a cognitive understanding of mathematics. After all, mathematics is the foundation of science and most science-related areas of learning. As a former professor of chemistry and chemical-physics related courses, much of the difficulties students experienced could be traced back to devoting considerable time to the contents in this book. In truth, it could also be of considerable value to other professions where mathematical formalisms are used, such as economics, political science, and business programs. Having students exposed to this learning at an early age would be extremely valuable to achieving a high retention of those interested in STEM-related professions. In this book, the Kinards delineate a mechanism for creating analytical-critical thinking and forming deep understanding of mathematics concepts. They show the systemic nature of rigorous mathematical thinking and conceptual learning from basic math through algebra and beyond. The lack of deep conceptual understanding of algebra causes many to give up on pursuing STEM education. After all, algebra is the gateway to higher mathematics and science. This book should be required as a basic competency for students interested in a STEM career, particularly since it is written in a fun and easy-to-understand manner. I highly recommend it! —William A. Guillory, PhD

Dr. Guillory, founder and president of Innovations International, is an authority on diversity, empowerment, leadership, creativity and innovation, and quantum-thinking. Prior to founding Innovations International Inc., Dr. Guillory was a physical chemist of international renown. He has lived, studied, and lectured in England, France, Germany, Japan, Switzerland, Poland, Mexico, Puerto Rico, Singapore, Hong Kong, and China. He has authored over 100 publications and several books on the application of lasers in chemistry and was chair of the Department of Chemistry at the University of Utah.

We dedicate this book to the memory of our beloved parents.

**Deacon Eddie Kinard and
Mrs. Corine Cooper Kinard**

**Mr. Samuel Gibson and
Mrs. Rosetta Gibson**

The Contents

Chapter 1: Context and Rationale for the Emergence of This Book 19

- Importance of Deep Mathematical Conceptual Understanding
- Roles of the Authors' Experiences and Spiritual Faith
- Early Judeo-Christian Mathematicians and Scientists Whose Paradigm was Theoretical Construction with Mental Experimentation
- African-American Female Mathematicians Whose Deep Conceptual Understanding Helped America Win the Space Race
- Laboratory of IDEAS

Chapter 2: Rigorous Mathematical Thinking: A Paradigm Shift in Mathematics Education... 49

- Need for Thinking in Mathematics Education
- Nature of the Mathematics Culture
- Mathematical Language and Tools of the Mathematics Culture
- Mathematical Knowledge
- Summary of Emergence of the RMT Paradigm

Chapter 3: Describing the Cognitive Function Model in Rigorous Mathematical Thinking 61

- Our Operational Definition of "Cognitive Function"
- Roles of Cognitive Tasks and Mediated Learning Experience
- The Anatomy of a Cognitive Function
- Feuerstein's Instrumental Enrichment Program as a Rich Source of General Psychological Tools

Chapter 4: Engaging Learners in Building Cognitive Functions in the Mathematics Learning Process.. 71

- Justifying and Validating Mathematical Outcomes
- Establishing a Cognitive Structure for Conceptual Understanding
- Developing Deep Conceptual Understanding of Quantity and Quantitative Relationships
- Forming Proportional Quantitative Relationships (Ratios, Rates, Proportions, Fractions)
- Forming a Functional Relationship between Two Variables

Acknowledgments

To our dear friend in Christ, Brother Eric Owens: Thank you for being a vigilant prayer warrior. You personify the man spoken of in James 5:16, "The effectual fervent prayer of a righteous man availeth much." And to Jennifer LuVert, owner of Forerunners Ink Writing & Design Services: Thank you for helping us prepare the book for publication.

FOREWORD

By Louis H. Falik, PhD and Sylvia Rousseau, EdD

This is a book about critical thinking. It is a book about the structural nature and development of thinking, and a unique perspective regarding how to achieve it. Critical thinking is replete with risks and dilemmas. For that reason, it is often paid lip service to rather than providing a functional strategy for implementation. This is particularly evident in the political spheres, witnessed by the "smoke screens" evidenced in the media and in many professional arenas around such hot-button topics as STEM curricula and the Common Core. My concern about the STEM and Common Core proposals is that they get bandied about in the media and common parlance to such an extent that they become almost meaningless—more catchwords and acronyms that no one really understands or thinks about. Ultimately, this book will provide the reader, and hopefully the decision-makers, with guidelines to move from lip service to application, to deeper knowledge and productive understandings. I will describe below why the Kinards are particularly well placed to offer this bridge to knowledge and action.

First, several comments regarding the dilemmas involved in a consideration of critical thinking. They are addressed in this book at several levels. For several decades, the importance of developing thinking skills has become a primary concern of educators and scholars concerned with the development of intelligence and both occupational and academic proficiency. This is reflected in a plethora of programs and curricula designed to foster thinking skills, and the agendas of professional organizations in education such as the Association for Supervision and Curriculum Development (ASCD). But the concern seems to manifest more in theory than in practice, and when evidenced in practice, there are significant impediments to true implementation. Several years ago, Professor Reuven Feuerstein (who is heavily attributed in this book) and I wrote a monograph addressing the issue, which was adapted for a journal article (Feuerstein and Falik, 2010). That work confronts the curricular issues involved in teaching thinking skills. This foreword gives me an opportunity to take the issue to a deeper level of consideration. Critical thinking is risky because of the political and academic dilemmas attendant to the development of thinking skills, particularly if they have the embedded quality of "criticality" as inherent objectives. And further, for the learner him or herself, there are risks!

Why is this so? Thinking is risky! To think (especially critically) involves challenges.

One may not know where one is going and to what destination will one arrive. One may arrive at totally unexpected outcomes. For established institutions, there are dangers. New knowledge has the potential of overturning or modifying that which is "known." One's conclusions may be at odds with conventional wisdom, or politically acceptable positions. One may challenge "expert" opinion or institutional arrangements. A poignant example is when the governmental position of "energy Tsar" was proposed during the Nixon administration, and the president was faced with selecting the first administrator of the new agency; he was given the choice of an academic (a scientist) or a politician. His advisors warned him against the scientist, "He will cost you a lot of money, he will propose innovative policies, etc." Better to select the politician, "He will be a safer choice, no new or costly ideas!"

And yet, we need critical thinking if we are to advance our social and technological agendas. The emphasis on the conceptual foundations of the Common Core proposals hold that learners advance through the acquisition and understanding of concepts that are then applied to specific knowledge. One does not move confidently or competently toward the STEM curricula (and end-results) without both affinity and understanding of the underlying concepts. So much of the "sciences" that frighten, or are seen as unapproachable by so many students, are ultimately accessible through the mastery of relatively simple concepts that are then amenable to elaboration into the "technicalities" of the sciences—from "concepts" to skills.

The authors have understood this. They come to their formulations as students of the work of Professor Feuerstein, implementing his theories and practices in structural cognitive modifiability (SCM). Their contribution is to focus on mathematical structures and formulae to achieve SCM. They hold that mathematics can be understood, manipulated, elaborated, and warmly engaged with if one is exposed to foundational concepts and actively experiences them in a meaningful, optimistic, yet demanding way. This last aspect is what is meant by a central operational Feuerstein formulation, that of "mediated learning experience" (MLE).

Here I must share a personal experience that exemplifies all of the above. I first encountered the application of the approach articulated in this book when I was invited to observe the senior author, Professor James Kinard, mediating a group of adult learners in a Federal government-funded program called The Environmental Remediation Technician Training Program. In this program, participants were selected to obtain training in order to qualify for jobs as environmental pollution inspectors and responders. For this they needed to learn aspects of chemistry, physics, problem solving, and other highly technical skills. However, the participants were recruited from very marginal populations—low income, educationally disadvantaged (many school drop outs), many who had been incarcerated, with histories of addiction, etc.

When I entered the classroom, I observed James working with the students on pages of the Feuerstein Instrumental Enrichment program. Concepts and strategies were being discussed; and problems from the pages were being solved, analyzed, then followed by wide-ranging discussions of implications of the solutions arrived at. Listening to the discussion—the questions being raised by the students, the responses and challenges of the instructor—without knowing the context, one could well assume a university classroom with extremely bright and competent learners.

Professor Kinard understood the necessity of this cognitive learning as a foundation for the acquisition of later specific technical knowledge, and he had advocated for it as a curricular component of the whole program. He was, in addition to being a Feuerstein trainer, a university Professor of Chemistry. Parenthetically, many other Feuerstein advocates and trainers have come from the disciplines of the sciences and engineering. Later, my positive observations were confirmed by research results on the outcomes of the program: high levels of mastery of the technical aspects of the training, and statistically significant pre- and post-test results on tests of cognitive ability. These results were later published in an Italian journal (Kinard and Falik, 1999) and in proceedings from an international conference on cognitive enrichment (Kinard, 2001).

Another implication from the above example has to do with the dilemma of predicted underachievement of students from poverty-stricken environments. Proponents of the human potential for cognitive modifiability do not accept this as an inevitable consequence of the condition. Professor Feuerstein understood this, as do the authors of this book. For this reason, the Kinards' theoretical and practical orientation is grounded in the application of Feuerstein's theory of structural cognitive modifiability (SCM) and application of mediated learning experience (MLE). Their application of these constructs with the populations with whom they work testifies to their belief and the outcomes of their actions.

As we focus on the intention of this book, we accept that all learning is essentially the response to a state of tension—that of not knowing, wanting to know, and searching for access. MLE creates the conditions so that the learner becomes amenable to the process of engaging in a search for solutions. The search involves mastering skills (that the tasks demand) and a willingness to engage in the search (to take risks, be open to process, see the personal value and avenues to the future). Perhaps a good metaphor comes from the *Wizard of Oz*, in Dorothy's willingness to follow the yellow brick road to seek solutions to her problems, a combination of optimism and skills! When we read the student reflections offered in the early pages of the book, we see this combination, framed with a sense of self-confidence.

Another observation of relevance stimulated by this book is the question of content and process in exposure and mastery of learning. This is a quality inherent in all learning experience. However, the question can be raised regarding sequence—what comes first? And whether they can or should be separated? The authors imply that learning occurs in "a laboratory context." That caused me much thought. What happens in a laboratory?

My first reaction is that this is where things happen, where actions are undertaken, studied, modified, and conclusions are drawn. It causes me to re-examine the relationship between content and process. If one has "big ideas," those "big ideas" need to be tested through processes of isolation, manipulation, and observation of consequences. This makes content/process a two-way street. "Learning by doing" leads potentially to new paradigms that further lend themselves to the process of exploring, manipulating, and so on. In this new paradigm, presumably evolving from the "laboratory" learning context, concepts become "real" and evolutionary. Ultimately, perhaps the ultimate consequence is that content is changed.

The content vehicle for the cognitive processes that the Kinards are working on is mathematics, but it could be experienced in any of the scientific domains (chemistry, physics, etc.). They view mathematics as a system for describing, predicting, and generalizing relationships. I had to ask myself why use mathematics as the vehicle? I began to think about "numeracy" in a comparable context to "literacy." For me, numeracy is much more than manipulating numbers. It is the thinking about the world in quantitative terms and providing a differential perspective on the world of stimuli, which operates on both concrete (with objects and events) and abstract (symbolic, relational, etc.) levels. The process goal might be described as moving from "calculation" to representation. The other content domains offer similar opportunities. Thus, the explication occurring in this book is to develop the role of mathematical thinking as the vehicle for the learner's developing an understanding of the "how, what, where" of the information being encountered. For the authors, the progression available in mathematics is to become analytical critical thinkers, using mathematical models to orient and understand, and then moving toward complex problem solving in the real world.

For this they impose another Feuerstein concept, that of the cognitive functions, and show how working with mathematical concepts and formulations affects the elaboration of essential cognitive functions. These functions can then be applied to different and various situations and problems. This in itself is a major contribution of this book—it is one of the best operational descriptions of the role that cognitive functions play in the processes of learning. It accomplishes that by integrating task dimensions and the cognitive functions, so that the relationship of the two points to how the learning dilemma is resolved.

A deeper meaning implicit in the conceptual development offered by the Kinards is that the major change in concepts is in the way that they transform from descriptive labels that organize the world into what Vygotsky termed "psychological tools." When this transformation occurs, thinking is altered—in time, space, and function. Thus, is content transformed into something active and meaningful.

I now conclude my reflections by describing the text that follows as a "very deep book." Not hard to read, but so embedded with meaningful conceptual formulations and analyses that it leads the reader to thought and gives indications of options for action.

Professor Falik is a senior scholar at the Feuerstein Institute in Jerusalem, Israel. Falik has been a co-author with Dr. Reuven Feuerstein on numerous books, monographs, and scholarly papers. He is Professor Emeritus, Department of Counseling, School of Education, at San Francisco State University.

***

Scholars and practitioners can count on Drs. James Kinard and Gwendolyn D. Gibson Kinard to remove the mystic that has needlessly grown up around teaching and learning mathematics, particularly in K -12 schools. They take us beyond the drudgery of memorizing algorithms and procedures, regimented cookbook applications, and routine problems that have little meaning for students besides just getting the assignment done. The effects of this long-standing traditional approach to teaching and learning mathematics has created an elitism that caters to the way only a few students in the United States leave our public schools with mathematical joy or knowledge. It has also led to the low standing of US students in mathematics knowledge and skills in comparison to other industrialized nations.

Traditional mathematics education in US schools has resulted in relatively few students advancing to upper levels of high school mathematics and an even smaller number being admitted to college eligible to matriculate in college-level mathematics courses. Typically, students in the US take the minimum number of required math courses for high school graduation with only a select group meeting minimum requirements for college admission. Two-year, as well as four-year, colleges cite ever increasing numbers of students requiring remedial mathematics courses upon entering their institutions. Students of color, particularly African American and Latino students, have experienced the most severe negative effects of this deeply entrenched way of teaching mathematics that runs counter to what we have learned in the past 50 years about how people learn.

Drs. Kinard, as both practitioners and researchers, call for "rigorous mathematical thinking" (RMT) as a necessity for improving mathematics learning in the United States. It acknowledges what years of science research has made apparent. The human mind is capable of cognitive functions, and through cognitive functions, learners construct knowledge and meaning. The insistence of Drs. Kinard on honoring students' cognitive functions brings to life longstanding, but often disregarded, principles of learning.

Ralph Tyler (1949, p. 63), consistent with the premise established by John Dewey, made the seemingly simple, but profound, declaration: "Learning is what the learner does." And what the learner does is "think" as Drs. Kinard demonstrate in their work. Paulo Freire reinforces this principle in his discrediting of the notion that learners can acquire knowledge by a "banking" method in which the teacher attempts to deposit information in the learner's head. The brain rejects such efforts and insists on thinking as a means for learning. Nothing—neither multiple routine tasks, nor voluminous assignments, nor repeated attempts at passing the same mathematics classes—will substitute for this basic function of the brain.

I applaud the work of Dr. James Kinard and Dr. Gwendolyn D. Gibson Kinard. I have seen their work promote the joy of learning and math identities among low income students in a school where math learning is assumed to be too difficult for most students.

This book is worth the time spent reading it and implementing its core principles.

Dr. Rousseau was a professor of clinical education and urban scholars for the USC Rossier School of Education, where she taught in the EdD program, focusing on instructional leadership, diversity, and organization in the K-12 concentration. She is former principal of Santa Monica High School, former Los Angeles Unified School District assistant superintendent of Secondary School Services and former superintendent of LAUSD Local District 7. She is Professor Emerita, Rossier School of Education, University of Southern California.

PREFACE

After you have read and digested this book, you may call yourself a mathematician. Did you ever think you would get here? But heed this warning! The path this book advocates for your arrival here is quite different from the familiar route of academic education and job training—the teach, practice, and test avenue. Oscar Wilde warned us: "Education is an admirable thing. But it is well to remember from time to time that nothing that is worth knowing can be taught" (Wilde, n.d.).

Is that the reason for the steady dimming of the bright flames of curiosity and eagerness that so many of our young children bring into the early grades? As time passes, many students shut down or even quit. Their love of learning dies. Is that why so many teacher-centered classrooms and job-training centers are filled with passive, disengaged, and lost students? Could it be the result of continually attempting to pour into students ready-made knowledge and inert facts? Henry B. Adams stated, "Nothing in education is as astonishing as the amount of ignorance it accumulates in the form of inert facts" (Adams, n.d.).

The path described in this book to reach your destination calls for engagement, thinking, and understanding. Engagement promotes thinking and thinking constructs understanding, which in turn, stimulates involvement, cultivates intrinsic motivation and fosters inner joy. Each human being has an awesome biological-neural apparatus, the human brain. In this regard, let us consider the nature and scope of the Internet. The Internet is a system composed of millions of computers around the world that are connected through a single fiber-optic network. Your brain is similar to the Internet, but it is extraordinarily more complex and sophisticated. A major difference is that your brain has up to 100 billion "computers" called neurons, connected together through a molecular biological network of axons and dendrites that would put AT&T to shame (Internet of the Mind, 2007-2017; Ghose, 2016).

Profound evidence of the awesome capacity of the human brain is presented in the recent book entitled *Hidden Figures: The American Dream and the Untold Story of the Black Women Who Helped Win the Space Race*, written by Margot Lee Shetterly, and in the movie that has its name and basis (Shetterly, 2016). Both depict the phenomenal true story of African American-women mathematicians who were "human computers" for the National Aeronautics and Space Administration (NASA) before there were machines driven by silicon-semiconductor technology. The details of Shetterly's diligent six-year research and writing project reveal that those brilliant mathematicians, who were college graduates, fully manifested the attributes that we specify in this book as academic rigor.

Academic rigor is a mindset for critical engagement and a state of vigilance that is driven by a strong, persistent, and inflexible desire to know and deeply understand. It compels intellectual diligence, critical inquiry, and intense searching for truth, addressing the need for perception and comprehension. Academic rigor has the quality of being relentless in the face of challenge and complexity, and having the motivation and self-discipline to persevere through goal-oriented struggles.

Those attributes were indeed prevalent in the character and disposition of each of these trailblazers. In addition, those tenacious ladies undertook a double-edged moral imperative—civil rights in segregated Jim Crow Virginia and equality for women—as they performed their work at NASA with commitment and intellectual diligence to help the United States win the space race against the Russians.

The path we are taking in this book will engage you in thinking, building academic rigor, and constructing deep mathematical conceptual understanding. This path is not about trying to pour inert facts and knowledge into learners. Instead, it is about nurturing learners to construct organic knowledge through diligent thinking. This book presents the RMT Laboratory strategy for developing rigorous mathematical thinking and constructing deep understanding of big ideas in mathematics.

This path has several milestones worth noting. The first is engaging you to connect with and consciously experience your intellectual self. Here we heed the advice of Kahlil Gibran: "The teacher who is indeed wise does not bid you to enter the house of his wisdom but rather leads you to the threshold of your mind" (Gibran, 1923, p. 56). Leading you to the threshold of your mind is both an act of belief on our part and a call for defiance on your part. This book is written with this foundational belief:

> In every human being—although sometimes very deeply embedded—lies a creative spirit, a quest for challenge, and a thirst for inquiry. The tunneling required to uncover these quiet and seemingly nonexistent diamonds may be deep and intense, but the persistent, goal-oriented struggles truly provide breakthroughs to in-expendable enrichments. (Kinard, J. 2000)

This book will show you one such path to reach these goals. Through this belief in your potential, we call on you to "Defy all Limits! Reclaim the Joy of Thinking and Learning!" The threshold of your mind leads to the Rigorous Mathematical Thinking Laboratory of Ideas. In this laboratory, we will mediate you—that is guide, shape, give scaffolds—to help you draw from your prior knowledge and experiences to proactively engage in the essential work of the laboratory. We will mediate you to construct, validate, apply, re-invent, and elaborate your own IDEAS by Investigating, Discovering, Experimenting, Analyzing, and Synthesizing. Ideas are constructed in this manner through rigorous thinking. In this thinking laboratory, ideas tend to feed off each other and spark a synergy of creative drive, which when formed within the context of a vision or a compelling cause, becomes a movement ready to express itself.

This book is based on the premise that effective mathematics education requires the acquisition of a new culture. Mathematics has its own culture (its values, history, tools, language, ways of doing things, etc.) that is distinctively different from the culture of every learner who enters the mathematics classroom. However, unlike other multicultural classrooms, here there are no learners belonging to the majority or dominant culture, because no learner, or teacher for that matter, can claim mathematics as his or her native culture. And this demands a paradigm shift in mathematics education. Why is this important?

It is through the mathematics culture that one learns "how to learn mathematics." In this book, you will experience how the RMT Laboratory model brings about this cultural transition and paradigm shift. Why is the United States trailing other countries in producing STEM professionals and workers? Is it because students are passing through mathematics classrooms, from elementary school through college, deprived of the mathematics culture?

You will discover from reading and comprehending this book how Reuven Feuerstein's theory of MLE promotes this cultural transition and provides the mechanism for engaging learners in the psychological tools of the laboratory to build cognitive functions. You will also experience how we mediate learners to acquire and internalize mathematically specific psychological tools, such as mathematical language, a number line, an equation, a table, and the x-y coordinate plane, based on their unique structure-function relationship. Finally, you will gain practice in applying cognitive functions, through the orchestration of mathematically specific psychological tools, to construct deep conceptual understanding and produce mathematical proficiency in basic math, fractions, ratios, rates, proportions, and linear equations and functions.

Therefore, reading and comprehending the contents of this book can impact your ability to understand and do mathematics, as well as supply you with a fresh and meaningful approach to understanding the need for and the relevance of mathematics in everyday living. If you are a parent who is searching for a strategy to support your child's mathematics education, the vignettes on mediating learner engagement and the learner artifacts may be of great benefit to you. If you are a mathematics classroom teacher, a teacher mentor, or a professor of pre-college teacher education, the instructional principles and learning methodologies elucidated in this work can inform your pedagogy and classroom practice. If you are a mathematics education researcher, a STEM educator, or a STEM-company manager/administrator, the theoretical underpinnings of the RMT paradigm discussed in this book along with the body of evidence from the applied practices may be of strong interest to you. Finally, if you are in public policy or education decision-making, this book could offer some thought-provoking insights that can contribute to your work.

It is very troubling to contemplate what this world would be like without the laboratory of ideas of some prominent mathematicians, scientists, educators, and so on. What if you refused to think at your highest level? You nor humanity can afford this choice. So, on with reading this book and THINKING!

Context and Rationale for the Emergence of This Book

Importance of Deep Mathematical Conceptual Understanding

This book demonstrates how Rigorous Mathematical Thinking (RMT) can be fostered by engaging you, the reader, in building cognitive functions and constructing and applying deep mathematical conceptual understanding. This raises the key question: Why is deep mathematical conceptual understanding so important? The answer is two-pronged; mathematics concepts are both systemic and theoretical in nature.

Regarding their systemic character, all mathematical ideas belong and contribute to a unified, coherent structure of conceptual meaning. For example, as we will illustrate later in this book, much of the conceptual nature of basic math, including the basic operations, fractions, ratios, rates, congruence, etc., along with the algebraic idea of slope in linear equations and functions, belongs to the systemic conceptual structure of forming proportional quantitative relationships. The nature of each of these topics, when acquired from the aspect of conceptual understanding, synthesizes a network of dynamic inter-relationships; this network enlarges one's mathematical comprehension and enhances his or her mathematical ability. When a learner constructs deep conceptual understanding of a big mathematical idea, he or she constructs a new pathway through which one's prior mathematical knowledge integrates into a larger more coherent body of meaning and builds a productive disposition toward complex problem solving. The learner who is relentless in pursuing the construction of deep understanding of mathematics concepts in an organized, progressive manner refuses to devote time to frantically preparing to take high stakes tests in mathematics, such as ACT (American College Testing), SAT (Scholastic Assessment Test), PARCC (Partnership for Assessment of Readiness for College and Careers), etc. The systemic nature of his or her conceptual understanding is progressively expanding to demonstrate strategic competence for any type of mathematics problem solving or assessment that is consistent with his or her exposure to the targeted concepts.

Now, we will focus on the theoretical nature of mathematics concepts and describe how it contributes to the importance of deep mathematical conceptual understanding. Theoretical understand-ing of an object or a process is derived from constructing its ideal form and being able to mentally experiment with it. A theoretical concept is both generative and universal. For example, the theoretical concept of a circle is produced by the rotation of a line segment, with one free endpoint and one fixed endpoint. The method for the theoretical construction of the concept of circle involves giving full cognitive attention to rotating the free endpoint of the line segment while mapping out the path it takes for one complete revolution on a two-dimensional plane. The learner can engage in a series of mental experiments by changing the length of the line segment and observing that a new circle is constructed that differs in the two-dimensional space it occupies, the distance across it, and the distance around it. Such a definition is generative because it provides a procedure or precise logical method for the generation of any circle. It is also universal because all possible circles can be generated in such a way (Kozulin, 1998, p. 55). Thus, theoretical comprehension of a concept equips a learner to apply that understanding in an unlimited number of situations and problems with different specifics and in varying contexts.

The traditional approach in most classrooms for introducing a learner to the idea of a circle involves comparing many round objects that the learner sees and is familiar with. This approach is product-oriented and only helps the learner to distinguish round objects. On the other hand, cognitively engaging a learner in the method of manipulating a rotating line segment to construct different circles leads him or her to acquire a skill of theoretical comprehension, which is process-oriented (Kozulin, 1998, p. 55). Thus, the major goal of RMT Laboratory is to engage learners in acquiring mastery of mathematics through constructing and applying deep conceptual understanding of its big ideas.

Roles of the Authors' Experiences and Spiritual Faith

The RMT paradigm and the design of RMT Laboratory have emerged from our personal and professional experiences and our spiritual faith and belief system. Let us elaborate further on what RMT Laboratory is and what it does.

Mathematics is the study of patterns and relationships. Mathematics concepts are derived from theoretical thinking regarding the relationships between patterns that exist within ordered-structures. RMT Laboratory is an analytical-critical thinking laboratory that teaches people to "think about their thinking" to learn how to learn mathematics with deep conceptual understanding and apply that understanding to chemistry, physics, and other areas. The Laboratory begins in the human mind and extends into the universe of natural phenomena and man-made activities, where patterns and relationships abound, ready for **I**nvestigating, **D**iscovering, **E**xperimenting, **A**nalyzing, and **S**ynthesizing into deep understandings.

The structures of patterns involve combinations of the concepts of time, space, matter, motion, and energy. Physics is the study of matter and its motion and behavior through space and time, along with related concepts of momentum, force, and energy. Chemistry is the study of the composition, structure, and properties of substances (matter) and the changes they undergo. In RMT, mathematical conceptual understanding drives chemistry, physics, and technological innovations and advancements.

A broad structure where patterns and relationships exist is within the phenomena of the electromagnetic (EM) radiation spectrum. EM is a form of energy that travels through space and is all around us. It consists of a combination of electrical and magnetic fields that are oscillating in planes that are perpendicular to each other and to the line of propagation or the direction of radiation. The EM radiation spectrum is the range of different forms of this energy, beginning from low energy (long wavelength) radio waves, and increasing in energy (decreasing in wavelength) to microwaves, infrared, visible light, ultraviolet, x-rays, and gamma rays (of high energy and short wavelength). Through mathematics and physics, we can form quantitative relationships among the concepts of time, space, motion, energy and matter for each segment of the EM radiation spectrum. Applications of deep mathematics-physics conceptual understanding, through engineering and technology, have brought about numerous benefits of EM radiation for human society.

It is our perspective that the mathematics of chemistry and physics is profoundly connected to our Judeo-Christian faith and belief system (private communications with Dr. Robert F. Kinard). James T. Kinard's purpose in life is to glorify God through the mathematics of chemistry and physics and help people construct and apply deep conceptual understanding of big mathematical ideas. Gwendolyn D. Gibson Kinard is committed to helping people recognize, realize, and value the beauty, the depth, and the power of their minds and their capacity to think.

Our cosmology is this: Jesus Christ, the Word of God (the Logos of God, the Eternal Self of God) is the thought of God. "If God is the thinker of the thought, James and Gwen Kinard can only be at optimal thought, which produces creativity and productivity, when we allow the Thinker of all thoughts to think His thoughts through us."[1] The Holy Scripture explicitly tells us three things concerning the nature of God

(Pink, 1930, p. 32). The first attribute: "God is spirit" (John 4:24), without an indefinite article. He is eternal and omnipresent, and He fills the heavens and the earth. He is beyond time and space. The second attribute: "God is light" (1 John 1:5). He is the essence and source of all spiritual light and is holiness, goodness, and excellency in His Person. The third attribute: "God is love" (1 John 4:8). He loves infinitely, and He is infinite and sovereign love Himself. The emergence of RMT through our combined perspective is specifically anchored in the following passages from the *Amplified Bible, Classic Edition*:

By the Word of the Lord were the heavens made, and all their hosts by the breath of His mouth. (Psalm 33:6)

In the beginning God (Elohim) created [by forming from nothing] the heavens and the earth. The earth was formless and void or a waste and emptiness, and darkness was upon the face of the deep [primeval ocean that covered the unformed earth]. The Spirit of God was moving (hovering, brooding) over the face of the waters. And God said, "Let there be light"; and there was light.(Genesis 1: 1-3)

We emphasize the connections between the concepts that are prevalent in mathematics, physics, and chemistry. The first

---

1 Dr. Ricky Allmon, private conversation

three verses in the Bible are paraphrased with the insertion of these concepts in brackets:[2]

> In the beginning [time] God (Elohim) created [by forming from nothing] the heavens [space] and the earth [matter]. The earth was formless and void or a waste and emptiness, and darkness was upon the face of the deep [primeval ocean that covered the unformed earth]. The Spirit of God was moving (hovering, brooding) [motion] over the face of the waters. And God said, "Let there be light"; and there was light [energy].

The Sovereign God created out of nothing the fundamental concepts for all mathematics, physics, and chemistry. God is spirit. "The Spirit of God was moving (hovering, brooding) over the face of the waters."

From the time of creation, constant reference is made in Hebrew Scripture to the Messiah and the Messianic hope of Israel (Parsons, n.d.). The Spirit of God (רוּחַ אֱלֹהִים), Rauch Elohim, was moving (hovering, brooding) over the face of the waters. The Hebrew word for move is *m'rahaphet*, and it also means to flutter or shake. We now turn to the metaphor presented by Moses in Deuteronomy 32:11 KJV: "As an eagle stirreth up her nest, fluttereth over her young, spreadeth abroad her wings, taketh them, beareth them on her wings . . ."

When a mother eagle or water fowl holds her slightly overarching and outstretched wings over her new brood, she slowly shakes, flutters, or trembles, stirring the air to maintain adequate distribution of heat and air for the well-being of the young eaglets or young chicks. When Jesus Christ, the Spirit of God, brooded, hovered, and fluttered over the face of the waters and the unformed earth, He was meticulously establishing the architecture of the laws of thermodynamics, along with the other principles of mathematics and science, to fashion, shape, and maintain an inhabitable and productive place for the prize of His creation, human beings. Thermodynamics is the physics that deals with the relationships and conversions between heat and other forms of energy (such as mechanical, electrical, and chemical energy). This is further articulated in the following Scriptures from the *Amplified Bible, Classic Edition*:

> For it was in Him that all things were created, in heaven and on earth, things seen and things unseen, whether thrones, dominions, rulers, or authorities; all things were created and exist through Him [by His service, intervention] and in and for Him. And He Himself existed before all things, and in

---

2 Private communication with Dr. Donnie Collins

Him all things consist (cohere, are held together). (Colossians 1:16-17)

He is the sole expression of the glory of God [the Light-being, the out-raying or radiance of the divine], and He is the perfect imprint and very image of [God's] nature, upholding and maintaining and guiding and propelling the universe by His mighty word of power. When He had by offering Himself accomplished our cleansing of sins and riddance of guilt, He sat down at the right hand of the divine Majesty on high. (Hebrews 1: 3-4)

We emphasized earlier this attribute of God's nature: God is light. Genesis 1:3 reads, *"And God said, Let there be light; and there was light."* When God said, "Let there be light," physical light came forth at 186,282.397 miles per second. Christ, the Word of God—the Logos of God— the spiritual Light of the world is the agency of God in all creation.

In the beginning [before all time] was the Word (Christ), and the Word was with God, and the Word was God Himself. He was present originally with God. All things were made and came into existence through Him; and without Him was not even one thing made that has come into being. In Him was Life, and the Life was the Light of men. And the Light shines on in the darkness, for the darkness has never overpowered it [put it out or absorbed it or appropriated it, and is unreceptive to it]. (John 1:1–5)

The immutable faithfulness of God, expressed in the conservation of principles and properties in the physical realm since the beginning of creation, enables humans to apply mathematical thinking to produce the scientific ingenuity and technology to derive benefits that sustain the quality of life. The following are some examples of the benefits derived from the EM radiation spectrum for human society.

Radio waves are used for communication through television, cellphones, and radios. Microwaves are used for cooking food and radar. They are also used by traffic speed cameras and in mobile phones. In some situations, microwaves can be utilized to treat health issues over drugs and serve as an alternative to surgery. For example, microwaves have been used to heat an enlarged prostate to reduce its size. Infrared radiation (IR) is in heat-sensitive thermal imaging cameras, which are used as night-vision cameras and in nocturnal animal research. IR is also extensively used in astronomical observations and in remote controls, which include DVD players, TV remotes, and projectors. The fields of optometry and opthamology have emerged through the interface of the physics of the visible region of the EM radiation spectrum with the structure, physics, biochemistry, and physiology of the human eye to detect and treat occular deterioration and diseases. Ultraviolet (UV) light stimulates production of vitamin D and has been utilized to treat diseases like rickets, psoriasis, eczema and jaundice. UV light has also been used in tanning beds, for hunting by illuminating blood trails from wounded animals, in forensic science to search for clues at crime scenes, and as a form of lasers to read information from the disc in CD/DVD players.

In addition, UV light serves as a tool to combat counterfeiting as ultraviolet verification is inserted as a marker on government and other sensitive documents and $20 bills. X-ray radiography is used to help detect or diagnose bone fractures, dental issues, some tumors, infections (such as pneumonia), kidney stones, heart problems (such as congestive heart failure), blood vessel blockages, digestive problems, and foreign objects. Gamma rays are used in the medical field to treat cancers and tumors, in PET scans and bone scans to detect malignancies, and in autoclaves to sterilize medical instruments. Gamma rays are also used in the food industry to irradiate food to kill bacteria and pathogens and to interrupt the ripening or sprouting stages to help maintain the freshness of fruit and vegetables. Gamma rays are also employed to inspect castings for cracks and defects.

God creates! Humans discover! As finite man discovers snippets of the Infinite God's expansive creation, without recalibrating his thinking, he may perceive that the physical universe is expanding.

Early Judeo-Christian Mathematicians and Scientists Whose Paradigm was Theoretical Construction with Mental Experimentation

Deep mathematical conceptual understanding is promoted through theoretical thinking. As discussed above, two key features of theoretical comprehension of an object are to construct its ideal form and mentally experiment with it. Many early Judeo-Christian mathematicians and scientists generated and validated knowledge through a paradigm of mental experimentation. Below, we present a brief overview of four of these theoretical thinkers: Nicholas Copernicus, Johannes Kepler, Galieo Galilei, and Blaise Pascal (Famous Scientists Who Believed in God, n.d.).

Nicholas Copernicus (1473 – 1543), a Polish astronomer, derived the first mathematical model that depicts the planets moving around the sun. He sometimes acknowledged God in his mathematics-scientific work. **Johannes Kepler** (1571 – 1630), an erudite mathematician and astronomer, studied light and formulated the laws of planetary motion about the sun. The field of astronomy shifted abruptly toward a modern perspective when he instituted the notion of force. Kepler, a devout Lutheran, elaborated in his scientific writing about how space and the heavenly bodies portray the Holy Trinity.

Some of these scholars were compelled to execute spiritual rigor when confronted with opposition and barriers that could have compromised their faith and their work. Spiritual rigor, like academic rigor, is a mindset for critical engagement and a state of vigilance that is driven by a strong, persistent, and inflexible desire to know and deeply understand. But unlike academic rigor, its state of vigilance is faith with hope, and its source of enlightenment (understanding) is the work of the Holy Spirit in reading and meditating on the Word of God. Also like academic rigor, spiritual rigor is the quality of being relentless in the face of challenge and complexity; but unlike academic rigor, it is inspiration along with motivation, and trust and confidence in God, along with self- discipline to persevere through goal-oriented struggles.

Galileo Galilei (1564 – 1642), an Italian mathematician, physicist, astronomer, and engineer, was confronted with professional antagonism from fellow scientists and ostracism from the Catholic Church. His controversial work on the solar system that was published in 1633 showed no substantial proof of a sun-centered structure and excluded the correct elliptical orbits of planets elucidated by Kepler some twenty-five years earlier. Galileo's staunch embracement of heliocentrism (the astronomical model in which Earth and other planets revolve around the Sun at the center of the Solar system) and Copernicanism was contradictory to his fellow astronomers who subscribed to either geocentrism or the Tychonic system.

Prior to this conflict, around 1591, Galileo demonstrated that the speed of a falling object is not proportional to its weight, as Aristotle had contended. From 1592 to 1630, he developed a telescope that facilitated his observation of lunar objects and craters, the four largest satellites of Jupiter, and the phases of Jupiter. In addition, he discovered that the Milky Way was composed of stars.

Ushered before the Roman Inquisition in 1615, Galileo was found "vehemently suspect of heresy" and was forced to spend the rest of his life under house arrest (Galileo, Astronomer, Scientist, n.d.; Galileo in Rome for Inquisition, n.d.). It was during that confinement that both his academic and spiritual rigor were pronouncedly manifested. He hunkered down and wrote two new sciences, which captured the work he had performed some forty years earlier by synthesizing his most impactful theoretical research, which was on dynamics. Galileo was among the first modern thinkers to clearly emphasize that the laws of nature are mathematical. Through his steadfast and unshakable faith in God, Galileo emphatically stated the Bible cannot err, and saw his sun-centered system as an alternate interpretation of the biblical texts (Famous Scientists Who Believed in God, n.d.)

However, over time, the Church recognized the truth in science (which was created by God) and lifted the ban on most of Galileo's works embracing Copernican theory in 1758. The Church had completely dropped its opposition to heliocentrism by 1835. But it wasn't until 1992, three hundred fifty-nine years after Galileo's conviction, when Pope Paul II of the Catholic Church acquitted him (Hellman, 1998).

Blaise Pascal (1623 – 1662), a French mathematician, physicist, inventor, and theologian, laid the foundation for the modern theory of probabilities. Around the age of ten he was conducting original investigations in mathematics and physical science. At age sixteen, he wrote an impressive treatise of projective geometry and commenced constructing calculating machines known as Pascal's Calculators. Later, he wrote expansively on mathematical theories and derived the principles of vacuums and the pressure of air (Blaise Pascal, n.d.). Although Pascal's triangle, an array of binomial coefficients, in the Western world was credited to Blaise Pascal, other mathematicians studied it centuries prior to him in India, Persia (Iran), China, Germany, and Italy (Pascal's triangle, n.d.).

Although Pascal was reared a Roman Catholic, at the age of thirty-one in 1654, he experienced a radical spiritual transformation when he had a vision of God. He turned completely from mathematical and scientific endeavors to theology. In his most prominent theological writing, the *Pensées* ("Thoughts"), he set forth a strong apologetics for Christianity, with a focus on the idea of Pascal's Wager. His last words were, "May God never abandon me."

During Pascal's staggering encounter in the presence of God, it was apparent that he wrote hasty notes during and/or following this overwhelming experience. After transcribing these notes on a piece of parchment, he sewed them into the lining of his coat.

From about half past ten in the evening until about half past twelve:
Fire! God of Abraham, God of Isaac, God of Jacob, Not of the philosophers and scholars. Certitude. Certitude. Feeling. Joy. Peace. God of Jesus Christ. "Thy God and my God." Forgetfulness of the world and of everything, except God. He is to be found only in the ways taught in the Gospel. Greatness of the Human Soul. "Righteous Father, the world hath not known Thee, But I have known Thee." Joy, joy, joy, tears of joy. I

have separated myself from Him. "They have forsaken Me, the fountain of living waters." "My God, wilt Thou leave me?" Let me not be separated from Him eternally.
"This is eternal life, That they might know Thee, the only true God, And Jesus Christ, whom Thou hast sent."
Jesus Christ.
Jesus Christ
I have separated myself from Him:
I have fled from Him,
denied Him,
crucified Him.
Let me never be separated from Him.
We keep hold of Him only by the ways taught in the Gospel.
Renunciation, total and sweet.
Total submission to Jesus Christ and to my director.
Eternally in joy for a day's training on earth.
"I will not forget thy words."
(Ps 119:16) Amen.

PRAYER (traditional language):
Almighty God, who didst grant to thy servant Blaise Pascal a Great intellect, that he might explore the mysteries of thy creation, and didst kindle in his heart a love for thee and a devotion to thy service: Mercifully grant to us thy servants, according to our several callings, gifts of excellence in body, mind, and will, and the grace to use them diligently and to thy glory, through Jesus Christ our Lord, who liveth and reigneth

with thee and the Holy Spirit, one God, now and for ever" (Kiefer, n.d.).

Thus, we have provided a brief overview of some early mathematicians and scientists who embraced theoretical conceptual thinking and whose worldview was that God is the Sovereign creator and maintainer of the universe.

African-American Female Mathematicians Whose Deep Conceptual Understanding Helped America Win the Space Race

Hidden Figures. Have you read the book? Have you seen the movie? This recent book and movie have provided an epiphany for readers and viewers that trashes assumptions of the prevailing culture about the abilities of females and blacks in mathematics and science. *Hidden Figures* is the phenomenal narrative of African-American female mathematicians at the vanguard of the women's and civil rights movement, whose deep mathematics conceptual understanding, along with their academic and spiritual rigor, helped launch America ahead of the Russians in the space race. Both the book, written by Margot Lee Shetterly (Shetterly, 2016), and the movie depict the marvelous true story of African-American women mathematicians who were "human computers" for the National Aeronautics and Space Administration (NASA) before there were machines driven by silicon-semiconductor technology.

Dorothy Vaughan, portrayed by Octavia Spencer in the movie, became supervisor of the all black, segregated West Area Computers ([A]Lee Shetterly, n.d.; Biography.com Editors, n.d.; Dorothy Vaughan, n.d.). During an interview, Spencer described the African-American female computers this way: "They were highly educated and they were moms and they were dreamers and they had fierce natures. And so there was so much about who they were that wasn't lost on me" (Blair, 2016). In another interview, Shetterly stated:

> When I think about these women . . . I think it's the everyday courage to be in a new situation where the expectations are very low, perhaps, and to stick with it and just, through force of will and through your own talent, decide that you're going to defy those expectations. That takes a lot of guts. It takes a lot of guts and a lot of gumption, and I feel like I find these women to be role models. I've learned so much from their stories. (Mirk, 2015)

Vaughan was among the first cohort of blacks to be employed as mathematicians at NASA. They were isolated in a segregated section and were responsible for performing the mathematical computations for the engineers conducting aeronautical experiments. Armed with the tools of slide rulers, calculators, and film readings, they generated the data the engineers needed to conduct various performance testing, such as the variables effecting drag and lift of the aircraft. Those experiments were conducted in the wind tunnels at Langley.

With the ingenuity and self-discipline to persevere through goal-oriented struggles, Dorothy Vaughan prepared in advance to strategically assist NASA in the transition from human to electronic computing. NASA was faced with a huge problem that demanded a faster method of calculating equations used for space exploration and travel. To assist in solving this dilemma, NASA acquired and installed the gigantic, room-size 7090, one of IBM's first transistor-based computers. No one from NASA's teams knew how to make it work. Vaughn, through her deep mathematical conceptual understanding and productive disposition, taught herself the programming language FORTRAN and successfully figured out how the system's logic operated. She trained her co-workers, the "West Area Computers" to program and operate the 7090. Her trailblazing efforts led her to be promoted as head of the programming section of the Analysis and Computation Division (ACD) at Langley and got the human computers new jobs. Mrs. Vaughan's Christian faith led to her extensive membership at St. Paul AME Church in Newport News, VA, and her active participation in the music department; the Dora Brown Missionary Society, and the Bread Distribution Program (Obituary: Dorothy Vaughan, 2008).

In 1951, Mary Jackson, played in the movie by Janelle Monáe, began working at the Langley Research Center as a research mathematician or computer in the West Area Computing division under the supervision of Dorothy Vaughan ([B]Shetterly, n.d.). She had earned a bachelor's degree in mathematics and physical science nine years earlier. Throughout her professional journey, she never recoiled in the face of challenge or difficulty. Equipped with a thirst for learning and a drive for scientific investigation, she accepted an offer by engineer Kazimierz Czarnecki to work in the Supersonic Pressure Tunnel, where she gained hands-on experience in studying forces on model designs with winds at almost twice the speed of sound.

Two years later, a major obstacle presented itself when she attempted to follow through on Czarnecki's suggestion that she prepare herself for promotion from mathematician to engineer. To achieve her goal, she was required to take graduate courses in mathematics and physics after work hours through the University of Virginia that were held at all-white Hampton High School. Unreservedly, she applied the self-discipline to persevere through her goal-oriented struggle by cleverly acquiring special permission from the city of Hampton officials. Jackson completed her courses and was promoted to aerospace engineer in 1958 as NASA's first black female engineer. She later worked as an engineer at NASA in the Compressibility Research Division, Full-Scale Research Division, High-Speed Aerodynamics Division, and the Subsonic-Transonic Aerodynamics Division. During her 34-year career at NASA, she achieved the most senior title within the engineering department (Mary Jackson, engineer, n.d.).

Mary Jackson was also a genuine STEM educator long before the acronym was coined. Following her graduation from college, she served as a mathematics teacher at a black school in Calvert County, Maryland. In the 1970s, she provided youngsters in the science club at Hampton's King Street Community center with the rich, firsthand experiences of building and carrying out scientific investigations with their own wind tunnel. She was cited in a local newspaper, stating, "We have to do something like this to get them interested in science. Sometimes they are not aware of the number of black scientists, and don't even know of the career opportunities until it is too late." Her commitment to Jesus Christ was expressed through her service at Bethel AME Church in Newport News, VA, and her extensive community service ([B]Shetterly, n.d.; Mary Winston Jackson Obituary, 2005).

Katherine Coleman Johnson's parents recognized that the youngest of their four children had a high aptitude for mathematics when she was at a very young age (Katherine Johnson: A Lifetime of STEM, 2013). Confronted with the obstacle that White Sulphur Springs, Greenbrier County, West Virginia, where they lived, did not provide public schooling for African-American students past the eighth grade, they took their children to Institute, West Virginia, to continue their education. The depth of the parents' commitment to their children getting the best education possible is evident in that the father remained in White Sulphur Springs to work year-round, while the mother and children resided in Institute during the school year and in White Sulphur Springs during the summer. Katherine was only ten years old when she enrolled in a laboratory high school located on the campus of West Virginia State College (now West Virginia State University), a historically black college (Katherine Johnson, n.d.; Shetterly, 2016; Loff, 2016).

Throughout her high school matriculation, the dedicated teaching and mentoring of Angie Turner King, a mathematician and chemist, provided a sanctuary of intellectual nurturing through which the progressive grounding of Katherine's mathematical conceptual understanding was promoted (Warren, 1999, pp. 148 - 150).

King, who was the first African-American woman to gain a PhD in mathematics, taught Katherine geometry, and Katherine thrived in mathematics throughout high school (Angie Turner King, n.d.; PhD degree to Mrs. King, 1955). King's PhD dissertation was on the analysis of early algebra textbooks used in American secondary schools before 1900 (King, 1955). King's exceptional quality of teaching and mentoring across her career established a legacy of helping prepare many students for advancing to post-graduate studies. During an interview, Johnson identified King as having major impact in her development. She stated that King was "a wonderful teacher—bright, caring, and very rigorous" (Warren, 1999, p.14).

Katherine's strong interest in astronomy germinated through conversations with her high school principal, who directed her attention to the stars and constellations, as they walked home together each evening from school (O'Conner and Robertson, 2016). Following her graduation from high school at age fourteen, Katherine was naturally compelled to attend West Virginia State College (WVSC), given that the high school was a laboratory of the college, located on its campus, and that she was awarded a full scholarship covering her tuition fees, and room and board. At WVSC, Katherine's understanding of the systemic nature of the mathematics discipline took root and her mathematical talent blossomed as a number of professors discovered her exceptional ability and drive and took her under their wings. Among those caring and visionary professors were Dr. King, who had mentored Katherine throughout high school, and the distinguished William Waldron Schieffelin Claytor, the third African American to earn a PhD in mathematics (Shetterly, 2016; William Claytor, a mathematical genius, n.d.; Collins, 2016).

Dr. Claytor, taking notice of Katherine's intellectual diligence and mindset for critical engagement, informed her, "You'd make a great research mathematician." And he committed to helping her become one. He insisted that she take all of the mathematics courses presented in the catalogue that were essential to pursing her goal. He even designed a class in analytic geometry of space just for her. Katherine graduated *summa cum laude* from college in 1937 at the age of eighteen with degrees in Mathematics and French.

About twenty-one years later and after working at NASA as a human computer under the supervision of Dorothy Vaughn, it was Katherine's academic rigor, deep conceptual understanding of mathematics and physics, and mastery of analytic geometry that affiliated her to the white all-male flight research team (Smith, 2015). Katherine Johnson, played by Taraji P. Henson in the movie, applied her mathematical and scientific

talents to support the successful flights of astronauts Alan Shepard, the first American in space, and John Glen, the first American to orbit Earth.

Scenes from the movie, with Henson portraying Johnson on a ladder fluently and methodically dispensing mathematical symbolism on a large chalkboard, implicitly demonstrate the hallmarks of the central principle presented in this book—that theoretical thinking and deep mathematical conceptual understanding drive strategic competence and computational proficiency in real-world problem solving. Pure math focuses on constructing and studying abstract concepts, and all mathematics concepts, unlike those of other academic disciplines, are theoretical in nature. And they are derived and studied through theoretical modes of thinking. Efficient, real-world problem solving demands theoretical thinking and the logical application of conceptual understanding. When asked about the reality conveyed in the movie, of the use of "old" math—Euler's method—to figure out how to get John Glenn back down from orbit, Johnson stated, "It seemed logical to me. I could see in my mind what I needed and sort of worked backwards." Working backwards is not just a matter of reversing procedure. Mathematical ideas are interrelated through a systemic conceptual network. Johnson's deep conceptual understanding established the logical pathway for her to work backwards.

Katherine Johnson's phenomenal contributions to the mathematics and science community and the United States went mostly unheralded until the age of ninety-six, when she was awarded the Medal of Freedom by President Barack Obama on November 24, 2015 (Former NASA Langley Mathematician to be Awarded Presidential Medal of Freedom Nov. 17, 2015). She was raised a Christian and grew up near St. James United Methodist Church (which was then an African American church) that she attended in White Sulphur Springs, West VA. Ms. Carolyn Bond, a church member recalled, "She definitely was known for counting her steps. She knew exactly how many steps it was from her house down Church Street to St. James Methodist Church" (Coble, 2017). Her ongoing commitment to Jesus Christ was evident in her hard work and dedicated service at Carver Memorial Presbyterian Church in Newport News, VA. There she served as a church officer and sang in the senior choir. A dedicated Presbyterian, she was a Commissioner to the 187th General Assembly, where she served on the finance committee She has been identified as a beacon of excellence (Sonia, 2017).

In this connection, we point out that John Glenn, whose re-entry trajectory was depended upon Katherine Johnson's mathematical thinking and accurate calculations, was also a devoted Presbyterian. He stated from the orbital view of his target planet: "Looking at the Earth from this vantage point, looking at this kind of creation and to not believe in God, to me, is impossible. To see (Earth) laid out like that only strengthens my belief " (Krohn, 2017).

Laboratory of IDEAS

In RMT Laboratory, we construct, validate, apply, re-invent, and elaborate IDEAS by **I**nvestigating, **D**iscovering, **E**xperimenting, **A**nalyzing, and **S**ynthesizing. Ideas constructed this way tend to feed off each other and spark a synergy of creative drive, which when formed within the context of a vision or a compelling cause, become a movement ready to express itself. And each idea is birthed through the formulation of good questions. Abraham Joshua Heschel stated,

> There are dead thoughts and there are living thoughts. A dead thought has been compared to a stone which one may plant in the soil. Nothing will come out. A living thought is like a seed. In the process of thinking, an answer without a question is devoid of life. It may enter the mind; it will not penetrate the soul. It may become a part of one's knowledge; it will not come forth as a creative force. (Heschel, 1955, pp.3–4)

In this regard, let us examine the laboratory of ideas of three notable individuals: a botanist, chemist, and agricultural scientist—**George Washington Carver**; a chemist and physicist—**Michael Faraday**; and a cognitive, clinical, and educational psychologist—**Reuven Feuerstein**. Let us consider these individuals and their laboratory of ideas in the context of three relevant factors: 1) the nature of, and relationship between, their theory and practice; 2) their source of inspiration; and, 3) a compelling cause through which the movement of their work expressed itself. This latter factor, a compelling cause, may have appeared within the context of a dilemma, a problem, a major issue or a challenge for others or society, but it tapped into the need system of the innovator and activated his intrinsic motivation. A compelling cause was the opportunity for the innovator to utilize his laboratory of ideas to address, "What must be done now!"

George Washington Carver is one of America's greatest scientists of the twentieth century and "an icon of American ingenuity and the transformative potential of education" (Bagley, 2013). He was born as a slave in Diamond Grove, MO, around 1864, near the end of the Civil War, in a small one-room log shanty. His life began within an awkward, turbulent context of geographical-sociological-political conditions. Although he was born about two years after the signing of the Emancipation Proclamation in 1863, he was still born into slavery (American Chemical Society National Historic Chemical Landmarks. George Washington Carver: Chemist, Teacher, n.d.). Missouri was not in rebellion because it never seceded from the Union and, thus, maintained slavery until the adoption of a new constitution on July 4, 1865 (Mc-Murry, 1981, pp. 9-10).

George's mother, Mary, was the property of Moses Carver, and his father, who was killed in an accident before George's birth, was the property of the owner of the adjacent plantation. After infant George and his mother were kidnapped by vigilantes during the troublesome Civil War times, Moses Carver was only successful in recovering the child. This left George, who was now without both of his natural parents, to be raised by Moses and his wife, Susan, on their farm in Newton County, Missouri.

George Washington Carver's laboratory of ideas began to take root at a very early age when he became intrigued by plants and began experimenting with natural pesticides, fungicides, and soil conditioners. He wrote in his autobiography that he was "a poor orphan who sought knowledge and hungered for scientific discovery" (Carver, A Brief Sketch of My Life, 1897). Applying his insights to improve the health of local farmers' garden plants, he was given the title "the plant doctor." Carver began his formal education at the age of eleven, when he left his surrogate parents' farm to attend a school for blacks, which was eight miles away located in the county seat of Neosho.

In Neosho, he was embraced by a set of black parents, Andrew and Mariah Watkins. Mariah was a midwife and nurse, with extensive knowledge of medicinal herbs, which appealed to young George. She was also very religious and developed a very nurturing and influential relationship with George and his young, inquiring mind. She stimulated him intellectually and spiritually, further grounding and reinforcing his faith and belief in divine revelation. George's deep faith in God and commitment to the Lord Jesus Christ energized him to struggle through severe poverty and deep racial discrimination to earn his master's degree in 1896, while he was part of the faculty at Iowa State University (George Washington Carver, n.d.). Professor Carver's laboratory of ideas expanded as he served as director of the Iowa State Experimental Station. Through his ongoing investigations, he discovered two types of fungi that were thereafter named after him. In April 1986, Booker T. Washington of Tuskegee Institute (now Tuskegee University), one of the first African-American colleges in the United States, invited this emerging renowned scientist to come to work at his institution. In his letter of invitation Booker T. Washington wrote:

I cannot offer you money, position or fame. The first two you have. The last from the position you now occupy you will no doubt achieve. These things I now ask you to give up. I offer you in their place: work —hard work, the task of bringing people from degradation, poverty and waste to full manhood. Your department exists only on paper and your laboratory will have to be in your head.

While Booker T. Washington was referring to the lack of a physical facility with the equipment, apparatus, and supplies fitting for a distinguished scientific researcher, Professor Carver was already well at home with his highly productive laboratory of ideas, already "in his head."

At Tuskegee, Carver's ideas were naturally forged into a ready-made context of compelling causes. His idea of crop rotation proved itself revolutionary in meeting the urgent financial needs for former slaves-turned-sharecroppers and poor white farmers who were experiencing reduced yields of cotton (Jones, 1975). Dr. Carver, a botanist, biochemist, and agricultural scientist, derived his revolutionary theory of crop rotation through diligent laboratory research and his ongoing study of the biological-physical universe and its natural laws. The principle underlying Carver's crop rotation strategy was to employ peanuts, which was both a simple crop to grow and had excellent nitrogen fixating properties, to enrich nitrogen-depleted soil from the practice of continuous cycles of growing cotton. Why was nitrogen fixation so essential to Dr. Carver's strategy of crop rotation?

First, studies had shown that nitrogen is essential for metabolic processes and growth in all forms of life. It is the central component of amino acids, which are the building blocks to chlorophyll in plants, and to proteins in animals. Although there is an abundance of atmospheric nitrogen in the ambient air, plants do not have the ability to utilize this form of the diatomic element. The most natural, least expensive, and most environmentally safe means by which plants can make atmospheric nitrogen biologically available for their use is through bacteria that convert the inert nitrogen to ammonium, which is the precursor for amino acids.

Plants of the legume family, such as peanuts, contain bacteria within nodules in their root systems, which produce nitrogen compounds that biochemically support the plants' metabolic and growth processes. At the death of these plants, the fixed nitrogen is released into the soil where it contributes to the life cycle of other plants and fertilizes the soil. Here we see evidence of Dr. Carver's theory and practice entering into dialogue and informing each other. A few years after going to Tuskegee, Professor Carver planted Spanish peanuts at the Agricultural Experimental Station, where he was director, and generated direct quantitative evidence that this legume restored usable nitrogen to starving soil. His 1905 Bulletin No. 6, "How to Build up Worn Out Soils" apparently was the result of knowledge gained from his theoretically inspired experiments (George Washington Carver, 1905, p. 4).

When confronted with the need to train and educate poor black and white famers of the practice of crop rotation and other techniques, Professor Carver invented the Jessup Wagon, a horse-drawn classroom and laboratory for demonstrating soil chemistry to those in isolated rural communities (Zabawa, 2008; Campbell, 1936; Felix, 1971, pp. 201-209).

Through his laboratory of ideas, Carver developed approximately 300 products from peanuts, patented three inventions, received an honorary doctorate, and made numerous other accomplishments. Dr. Carver stated:

> It is not the style of clothes one wears, neither the kind of automobile one drives, nor the amount of money one has in the bank that counts. These mean nothing. It is simply service that measures success. (Kremer, 1987, p. 17)

Professor George Washington Carver was a deeply devout Christian and a very humble man. His laboratory of ideas found many compelling causes that enhanced the quality of human life for people around the world. Overman wrote: "Carver was in full-time ministry. Not as pastor, but as a botanist-chemist." (Overman, 2010). For this dedicated scientist, God in the person of Jesus Christ is the Sovereign first cause of everything. Carver stated, "Never since have I been without this consciousness of the Creator speaking to me…. The out of doors has been to me more and more a great cathedral in which God could be continuously spoken to and heard from." (Dao, 2008). Carver is also quoted as saying, "I never have to grope for methods. The method is revealed at the moment I am inspired to create something new. Without God to draw aside the curtain I would be helpless" (Federer, 1994, p. 96).

The laboratory of ideas of Michael Faraday was also birthed at an early age. Both a physicist and a chemist, Michael was one of the most prolific scientists of the nineteenth century (Lamont, 1990). He was born on September 22, 1791 in the village of Newington, Sussex, England, to a poor family, during the tumultuous period of the French Revolution. At the age of thirteen, after receiving just a basic education at a local school, Michael was compelled by family poverty to work as a delivery boy for a bookshop. Moved by the youngster's work ethics, his employer promoted him to become an apprentice bookbinder in 1805 (Michael Faraday, 2014).

Over the next seven years in that position, he performed his work diligently, and during his free time, he read extensively, cultivating a passion for science, and particularly electricity. Three books sparked his interest above all. The first was Isaac Watts' *The Improvement of the Mind*, which contributed to nurturing his early Christian thought and providing spiritual inspiration for his inquiring mind. Dr. Watts' writing was also pedagogical, equipping young Michael, who was already giving careful attention to details, with a methodology to become organized and systematic in his thinking, notetaking, and work. The other two books were *The Encyclopedia Britannica*, from which he extracted knowledge about electricity, etc., and Jane Marcet's *Conversations on Chemistry*. The knowledge and intellectual intrigue gained from his reading propelled him to use some of his precious but sparse earnings to purchase apparatus and chemicals to test and validate ideas through empirical investigations. Here his deep interest in the concept of energy, particularly force, was seeded (Williams, n.d.).

Through that type of extensive reading, thinking, and experimentation, Faraday continued to construct knowledge, understanding, and new insights that chartered the path for his profound contributions to the fields of electromagnetism and electrochemistry. As a brilliant thinker and gifted experimentalist, Faraday conducted much of his research in the nineteenth century during the latter phase of the chemical revolution. Through his laboratory of ideas, Michael Faraday discovered complex phenomena that have practical, understandable qualities that effect everyday living. Some of these include electromagnetic rotation (1821); gas liquefaction and refrigeration (1823); benzene (1825); electromagnetic induction (1831); the quantitative relationships between chemistry and electricity as depicted through Faraday's law(s) of electrolysis (1833); the Faraday effect—a magnetic optical effect (1845); and diamagnetism as a property of all material (1845). (See Michael Faraday, 2014.) Faraday's 1852 paper, entitled "The Physical Character of Lines of Force," was a confluence of his prolific experimental knowledge and emerging theoretical and philosophical thought, which became a defining cornerstone of his legacy. James Clerk Maxwell took Faraday's work and others and formulated the mathematical apparatus, in a set of equations, that became the basis of all modern theories of electromagnetic radiation phenomena (Campbell and Garnett, 1882, pp. 513-556; Faraday's Line of Force and Maxwell's Theory of the Electromagnetic Field, n.d.).

Faraday was also a devout Christian whose theoretical principles and scientific researches were not only in harmony with, but also shaped by, his theology (Eichman, 1991; 1993, pp. 92-95). In an exhortation presented in London on July 7, 1861, Faraday expounded upon God's moral law, stressing that salvation cannot be acquired by the keeping of the law, for perfection can only come through Jesus Christ, Who lived a perfect life. He stated, "The law of God required perfect obedience, which man could not render, and it was in the room and stead of guilty man that Christ fulfilled it." He closed his exhortation with this:

And therefore, brethren, we ought to value the privilege of knowing God's truth far beyond anything we can have in this world. The more we see the perfection of God's law fulfilled in Christ, the more we ought to thank God for His unspeakable gift. (Selected Exhortations, 1991)

The renowned clinical, developmental, and cognitive psychologist, Reuven Feuerstein, revolutionized the fields of cognition and learning with his comprehensive theories of structural cognitive modifiability (SCM) and mediated learning experience (MLE), and his applied systems of learning propensity assessment device (LPAD), Feuerstein's instrumental enrichment (FIE) program, and shaping of modifying environments (SME) (Feuerstein, 1990; Feuerstein, Klein, and Tannenbaum, 1991; Feuerstein, R., Feuerstein, S., Falik, and Rand, 1979, 2002; Feuerstein, Rand, Hoffman, and Miller, 1980, 2006). Born with humble beginnings on August 21, 1921 in Botosani, Romania, he was one of nine siblings. While Professor Feuerstein's laboratory of ideas began emerging at the very early age of three, when he started reading and writing both in Hebrew and Yiddish, the exact date of his birth is shaped by the constructs of one of his own theories.

Feuerstein insisted that mediated learning experience (MLE) is the dynamic that perpetuates one's culture from one generation to another. Use of the Jewish calendar and the custom of dating by the nearest Shabat that one was born created uncertainty of Reuven's date of birth. Falik (in press, p. 71) wrote:

> So he dates his birth, and celebrated his birthday according to when Shabbat Nachamuz occurred, and that can vary as much as three weeks according to the Georgian calendar. The year is also in some dispute. At the celebration of his 80th birthday, his wife Berta whispered to me, so as not to spoil the event, that it might actually be his 81st, they were not absolutely sure. He actually chose August 22 as his "Birthdate."

Perhaps the birth of Dr. Feuerstein's lifelong passion for teaching, or should we say mediating, occurred at six years of age, when he took the audacity to help his playmates learn to read and write. At the age of eight, he embarked on his first experience engaging the learning disabled. Young Feuerstein was approached by a local elderly man to teach his 15-year-old mischievous son how to pronounce Kaddish. Reflecting on this experience Reuven stated: "To help the father die happily, I took on the challenge. I found the key to unlock the boy's intellect. Today he is 84, with a slew of grandchildren and great-grandchildren" (Ginsberg, 2005).

From his childhood through his young adult life, Feuerstein and his close-knitted family were forced to navigate through a social and political matrix in Romania that was shaped by the growing expression of anti-Semitism and the appeal of fascism that commenced around the late 1920s. Both conditions promoted the formation and success of the Iron Guard. The Iron Guard was fixed on the perception that the economic struggles for non-Jewish Romanians stemmed from the Jews having "colonized" Romania. Thus, driving the Jews out of Romania was their solution to this perceived problem. These conditions, coupled with the emergence of Nazi Germany as a European power, cultivated a festering deep, dark atmosphere of racial discrimination as the norm. Around 1937, the fascist government of Romania had morphed into a strong resemblance of Nazi Germany with comparable anti-Jewish laws that promoted Hitler's strategy to build one gigantic eastern front against the Soviet Union (Reuven Feuerstein, n.d.). The insidious nature of racism shrouds a toxic web of fear, hypocrisy (deception), and hatred—described by the great theologian Howard Thurman as the three hounds of hell—a cancerous agony within the souls of the perpetrators (Thurman, 1996). This troublesome context created a compelling cause for Reuven Feuerstein's laboratory of ideas, even as it was transitioning from infancy into its defining structures. It took place when Feuerstein was studying psychology at the University of Bucharest, that he had the need to join with others to help children whose parents were snatched away by the Nazis. Feuerstein was quoted as saying: "Afterwards, I heard that the school director told others, 'Reuven has a good heart, but he's stupid. Why is he teaching these poor children to sing, to think and to express themselves? He should be teaching them how to use a hammer and nails. That's what they need to know!'" (Falik, in press, p. 86; Ginsberg, 2005). Feuerstein's driving belief in human modifiability led him, some fifty years later, to co-author the book *Don't Accept Me As I Am: Helping Retarded Performers Excel* (Feuerstein, Rand, and Rynders, 1988).

A central feature of Feuerstein's theory of MLE is the role of culture in cognitive development. The first step in defining this role is to differentiate between cultural difference and cultural deprivation. Feuerstein insists that each culture has its own MLE based-systems for transmitting the culture from one generation to another (Feuerstein, Klein, and Tannenbaum, 1991; Feuerstein, Rand, Hoffman, and Miller, 1980, 2004; Kozulin 2001):

Individuals receiving sufficient MLE in their native culture have developed adequate learning potential to adapt and function in a different culture. On the other hand, individuals who are deprived of MLE in their own culture demonstrate a reduced learning potential that impedes their successful transition to a new culture and educational programs. It is not that these individuals' culture is depriving, but that these individuals are deprived of their culture because of a range of conditions, such as poverty, isolation from parents, disruptions such as violence, war, etc. In this regard, I turn to the effort of transitioning Ethiopian Jews, descendants of the Beta Israel communities of Ethiopia, into the modern society of Israel.

Feuerstein asserted that an immigrant group's ability to preserve cultural transmission is the essential dynamic for the process of adaptation to a new culture, rather than the difference or the "distance" between the original and the new culture (Kozulin, 2001, p. 3). When anchored in their cultural systems and processes, individuals in the immigrant group bring the experience of cultural learning and a strong sense of cultural identity, even though the content of their native culture and methods of transmission may be vastly different from those of the new culture.

The Ethiopian Jews lived in rather isolated rural communities. The mode of transmission of the native culture was oral and driven mainly by observation and imitation. Kozulin (2001, p. 7) reported:

> The older males sit in a group drinking coffee while narrating stories about the important events and heroes of the past. The narratives include highly elaborate word play, verbal riddles and creative poetic comments offered by the participants. Children who sit patiently and silently on the periphery of the story-telling circle gradually absorb the cultural content and verbal technique. Each such "session" lasts for hours and constitutes an integral element of everyday life. (Kozulin, 2001, p. 7)

Professor Feuerstein, reflecting on the cultural transmission of Ethiopian Jews into the modern technological society of Israel, said:

If you ask me why can we take Ethiopian people who are so culturally different that they function at the 5th percentile on the Raven (a cognitive assessment tool), and you give them just a little bit of mediation, and they grow to 60th and 70th percentiles, this shows the power of MLE. From this point of view, it also shows that these Ethiopian students were not culturally deprived, they were culturally different. MLE explains in the best way the difference between cultural difference and cultural deprivation. Cultural difference describes the person who has had mediation, but he must learn the new culture into which he has been transplanted, this is why he is different, but he has the tools to do it. Cultural deprivation describes someone who has not had access to the culture of his people or community, he has not been mediated, and is unlikely to have the tools to adapt—without mediation. (Falik, in press, p. 167)

From the age of three, Reuven devoted much time and attention to the constitution and study of Judaism. His theories, concepts, and practices are integrally and intimately rooted in his Judaic faith and life experiences. Professor Louis H. Falik, one of the writers of the foreword to this book, meticulously embellished on the Judaic roots of Feuerstein's theories and practices and the emergence of mediation in the Bible (Falik, in press, pp. 168 – 169). Dr. Falik quotes Professor Feuerstein:

> What I am telling you now comes from the time when I started to formulate my theory of MLE. And to fully understand it, you must promise me to read through the chapter of Bereshit (Genesis), and be intrigued by the lack of mediation. I myself cannot understand it. From a biblical perspective, the study of Noah, and the transformation—a transition with regard to the way culture is transmitted. Before, G-d did and people acted. After Noah, man works to understand and interpret experiences, and this is the beginning of mediation.... But you must understand, I bring the story of Noah as very strong evidence that the lack of MLE was not a casual thing … it was not referred to. Maybe it existed, but it was not explicitly spelled out. So the Torah does not tell you everything that has happened. Does it mean it didn't happen? But if you have a very clear event, one that makes it explicit, the Ark of Noah, to alert people, after 500 years, it was meant to serve as a message.

Perhaps the most compelling cause for both Professor Feuerstein's spiritual faith and theories and practices expressed itself just after he fled for his life from Romania to Palestine in 1944 and began his life's work with children of the Holocaust. Seeing the vicious atrocities of this grave human tragedy encapsulated in the traumas and handicaps of its young survivors created a deep and unprecedented need that charged him spiritually and intellectually. This compelling cause became the catalyst that ushered him and his team into serious investigation, discovery, experimentation, and analysis that evolved into the synthesis, crystallization, and formulation of his big ideas of "deficient cognitive functions," MLE, SCM, LPAD, FIE, Cognitive Map, etc. And it was driven by his faith and love—his belief that the image of G-d was within each and every one of these slow and often nonresponsive learners and that they can change, that they are modifiable. Due to the separation from their parents and family, and isolation from cultural nourishing, they were deprived of their native culture. They needed to be mediated.

The application of MLE brings about the establishment, restoration, and/or rehabilitation of cognitive functions. The Learning Propensity Assessment Device (LPAD) uncovers and defines the nature of "deficient cognitive functions" within the learner and scientifically designs a protocol for the amounts and types of mediation needed. The Feuerstein Instrumental Enrichment (FIE) program comprises the set of tools by which MLE is brought to the learner to achieve Structural Cognitive Modifiability (SCM).

Further evidence of the Judaic roots of Professor Feuerstein's practice is explicitly presented in the introduction and overview of the instrument Categorization of the FIE program. Mediating learners through the tools in this instrument helps them develop the strategies and mental processes for the organization of data in their "universe" into superordinate categories. In the unit on categorization according to subject and principle, Professor Feuerstein wrote the following (see Feuerstein and Hoffman, 1995, p. 8):

This can be likened to the position attributed to Adam by the Bible in his act of labeling the creatures of the world:

And out of the ground the Lord God formed every beast of the field and every fowl of the air, and brought them unto the man to see what he would call them; and whatsoever the man would call every living creature, that was to be the name thereof. And the man gave name to all cattle, and to the fowl of the air, and to every beast of the field.
(Genesis 2:19 – 20).

Through the act of labeling and classifying, Adam gained mastery of the creatures of the universe.

The theories and applied systems of Reuven Feuerstein's laboratory of ideas have fed off each other and have created a movement that continues to express itself in both clinical and academic settings around the world, and the lives of thousands of learners continue to be transformed.

Of course, I (the first author) did not personally know Dr. Michael Faraday. He lived a century and a half before my time. Nor did I personally meet Dr. George Washington Carver. He died about four months before I was born. But I have been blessed by Almighty God to sit at the feet (under the tutelage) of Dr. Reuven Feuerstein, serve on his international faculty, and practice his work in inner-city communities across the United States. I often greeted him privately and affectionately as "Thou son of Abraham." I remember a particularly nostalgic moment following a meeting we had in his hotel room at the Shoresh International Workshops held in the Judean Hills in the late 1990s. We were brainstorming about a technical issue, and I was somewhat perplexed because we had not resolved the matter before we stepped into the hallway to attend the next lecture. As we continued our conversation while walking down the hall, Professor Feuerstein, realizing my concern, put one hand on my shoulder and spontaneously began singing, with somewhat of a cadence as we walked, *We shall overcome, we shall overcome, we shall overcome someday; Oh, deep in my heart, I do believe, we shall overcome someday.*

On several occasions, I was invited to Professor Feuerstein's home in Jerusalem for a Shabbat meal. Each time Reuven led a hymn, with his melodious baritone voice saturating the atmosphere, I was inspired with rich, vibrant memories of my father, Deacon Eddie Kinard, leading hymns and songs at prayer services in our small, rural Baptist church in Irmo, SC, when I was just a child. Several times during other Jewish celebrations, Professor Feuerstein and his beloved son, Rabbi Rafi Feuerstein, requested that I lead the group in singing the Negro Spiritual "Go Down Moses." As Reuven's and Rafi's voices merged into harmony with mine, everyone joined in, and we experienced great jubilation and rejoicing. These were indeed rich moments of deep cross-cultural meaning, with Feuerstein's theory of MLE in full operation. All of us were anchored in Exodus 8:1: *"And the LORD spake unto Moses, Go unto Pharaoh, and say unto him, Thus saith the LORD, Let my people go, that they may serve me."* Personally, "my people" are both Israel and African American slaves with Pharoah representing the slave master. Today, "my people" are those who claim that they cannot do mathematics, and Pharoah is the entrenched practice of attempting to learn and apply mathematics without thinking and building understanding.

Each mathematician and scientist discussed in this chapter lived their story. From each story emerged a unique voice that, respectively in its era, contributed to the development of current-day mathematics and science disciplines and practices.

The lead author (James T. Kinard Sr.) and the co-author (Gwendolyn D. Gibson Kinard) have experienced individually and as collaborators, the power and necessity of recognizing one's story as a critical element in the projecting of one's voice. The power of an individual's voice is reflected in a person's drive and ability to create and contribute to the well-being of self, family and community. It is from this perspective that we share our individual and collective stories as mathematicians and scientists.

I (the first author) was born during the days of Jim Crow in rural South Carolina, where the ideals of Strom Thurman and stark racism permeated every fabric of political, social, and economic life. My father and mother, who made mathematics very appealing to me before I started first grade, with great intentionality mediated my two siblings and me into our Judeo-Christian faith from early childhood. They also mediated us in cultivating a positive self-image, with admiration for our African-American heritage, knowledge of our history, and love for our people. We read everything we got our hands on.

My dad was a gifted cabinet maker and skilled carpenter. People from far and near sought him out for his craftsmanship and specially designed cabinets. As soon as my brother and I were old enough to assist him, our father demanded precision in every measurement, cutting every board, and our giving full attention to details, even when sweeping up the trash. My mother, who only had a fifth-grade education, was talented in mathematics. Individuals from all walks of life, including college students and teachers, came to her to be tutored in mathematics. I can only imagine how far she would have gone if she could have gotten a full formal education. Perhaps a research mathematician!

The day we got electricity connected to our small country house, I saw the rooms light up from the ceiling to the floor. All before we had been reading by the dim flickering light from the kerosene wick lamps and candles. I was very excited. It seemed that things were transformed. Small, nearly invisible objects were suddenly distinct and clear. And we could read now with a greater amount of focus and clarity. It seemed that the words on the page came alive. It wasn't until I was in my high school physics class that I was able to more precisely quantify the difference in brightness through the unit of the lumen. A kerosene lamp or a candle gives off about eleven lumens, while a 100-watt incandescent light bulb provides 1,600 lumens. This is one piece of evidence of how Dr. Michael Faraday's laboratory of ideas profoundly impacted my young life. His discoveries of the basic principles of electricity and electromagnetic induction, along with the work of others, fueled the science, mathematics, and technology for household electric lighting, and this opened up a new world for me.

My interest in chemistry was solidified when my sister, Wilhelmina, who is four years older than me, was taking chemistry in high school. I covertly read her chemistry workbook, giving much attention to her detailed notes and the teacher's comments. My sister went on to become a registered nurse in October 1960. My brother, Robert, who became a high school art teacher, and later a Baptist minister, received his PhD His dissertation was entitled "Mediated Learning Experience in the Art Classroom for Cognitive Development."

The co-author, Gwendolyn, began life in Chicago, Illinois, then a rapidly expanding "urban oasis" for the "second wave" of newly migrated Negroes from southern states, including her parents, Samuel and Rosetta Gibson. She and her thirteen siblings (two died as infants) grew up in Englewood, a southside neighborhood and subsequently moved to the newly developed Cabrini-Green Housing Development on Chicago's near north side. She attended elementary, junior high, and her freshman year of high school on

Chicago's near north side. One day, her father made a surprise announcement that the family was moving into a new home in Gary, Indiana. So, in February, before the summer of the first round of riots during the mid-1960s, the Gibson Family settled into their new environment.

Gwendolyn's father had been raised as a sharecropper on a plantation in rural Mississippi. As a Negro, he was not allowed to go to school, which he desperately wanted to attend. He eventually did learn to read, write and study mathematics, all of which he loved, while serving in the US Army during World War II.

From her father, Gwen inherited a thirst for academic studies and became a lifelong learner. From her mother, she acquired the desire to learn and understand more about the relationship between God and His unique Son, Jesus, and why studying the Bible was important in one's life.

As a child, Gwendolyn loved nature and going to the public library to check out books. She enjoyed reading about birds, animals, and the science of human body systems. During the summer, she participated in a Bible camp in Southeast Michigan. Camp activities included nature study—identifying and studying plants, small mammals, reptiles and insects. She also learned techniques for building a campfire, engaged in various sports, and learned the importance of teamwork with camp leaders and fellow campers.

Those activities helped Gwen to increase her interest in studying mathematics and science. In elementary and secondary school, Gwen discovered two opposing realities that would challenge her for decades to come: 1) Not only did she have a strong interest in mathematics and science, but also learning algorithms and formulas was easy to comprehend; and 2) she was not challenged to learn and apply the concepts, particularly in mathematics, due to the emphasis on traditional textbook instruction. However, she persisted and completed a college degree in chemistry.

The lack of emphasis on conceptual understanding in teaching and learning still plagued her until she began her PhD dissertation research on high school science classrooms. During this time, she discovered conceptual change theory. Little did she realize that coupled with her educational psychology and special education courses in graduate school, that she was being prepared to encounter a transformational experience that would bridge the gaps in her mathematics and science learning.

While working as a district-level administrator with the Chicago Public Schools Systemic Initiative for Standards-based Mathematics, Science and Technology (MST), Dr. Gwendolyn Gibson was faced with a huge challenge. Part of her job as a facilitator and advocate for special education teachers and students in the MST Initiative was to identify and develop ways that provided students with disabilities full access to standards-based instruction. She recognized that more had to be created to ensure that both special educators and their students would be effectively included in the process. A question that burned within her was this: Is there a natural connection to learning mathematics and science for all students?

About a year later, while attending a National Science Teachers Association Conference, Dr. Gibson was introduced to her answer. It came in the form of an introduction to Professor Reuven Feuerstein's Instrumental Enrichment with Mediated Learning Experience (FIE/MLE). This introduction led to her studies with Professor Feuerstein, as well as to meeting her future husband, Dr. James T. Kinard Sr.

We have been blessed by Almighty God to be students of the Dr. Reuven Feuerstein, serve on his international faculty, and practice his work in inner-city communities across the United States. The contribution of Professor Feuerstein's theories and practices is invaluable to the creation of the Rigorous Mathematical Thinking paradigm and story.

It is very troubling to contemplate what this world would be like without the laboratory of ideas of George Washington Carver or Michael Faraday or Reuven Feuerstein. What if George or Michael or Reuven had refused to think at his highest level? And consider the human computers and their contributions to society. What if they had refused to think? This is something the world could not afford. And neither you nor humanity can afford for you to not THINK! ■

Chapter Two

Rigorous Mathematical Thinking: A Paradigm Shift in Mathematics Education

In This Chapter

- Need for Thinking in Mathematics Education
- Nature of the Mathematics Culture
- Mathematical Language and Tools of the Mathematics Culture
- Mathematical Knowledge
- Summary of Emergence of the RMT Paradigm

Need for Thinking in Mathematics Education

We begin with the following quote: **"Math illiteracy affects 7 out of every 5 people."** This quote expresses the deep need for mathematical literacy in our expanding technological world. But it also speaks to an even broader and pervasive need that is prevalent across the entire fabric of human society today: the severe poverty in the quality of our thinking in general. Doing anything without thinking can be costly and have dangerous consequences.

Evidence of this profound need has been demonstrated and documented in everyday living, along with places and institutions such as schools, the workplace, legislatures, courts, the body of politics, law enforcement, homes, etc. And it impacts all of us! Throwing around catchy phrases and buzzwords like "critical thinking," "rigorous tasks," "real-world problem solving," "higher-order thinking," "grit," etc., has not, within itself, met this need. Neither has technological advancement in the form of cellphones, calculators, computers, iPads, electronic games, etc. (see Delvin, 2017).

Thinking is truly a human endeavor, and it demands a serious investment of time and effort. Technology, regardless of the conveniences it brings, cannot think and learn for us. Neither can technology generate the authentic joy and feelings of competence that we, as human beings, derive from thinking and learning. However, a large segment of the human population regards thinking as both difficult and agonizing.

Thomas A. Edison (2016) stated the following bold statistics: "Five percent of the people think; ten percent of the people think they think; and the other eighty-five percent would rather die than think." Dr. Martin Luther King Jr. and Henry Ford suggested the reason for such discrepancy. King (2016) said:

> Rarely do we find men who willingly engage in hard, solid thinking. There is an almost universal quest for easy answers and half-baked solutions. Nothing pains some people more than having to think.

Ford (2016) stated:

> Thinking is the hardest work there is, which is probably the reason why so few engage in it.

A major claim in our Rigorous Mathematical Thinking paradigm is that learners' difficulties with mathematical tasks stem not from the lack of specific mathematical knowledge but from the absence of cognitive processes of a more general nature (Kinard and Kozulin, 2008, p. 81). In classrooms across the United States, students are required to learn mathematics by memorizing algorithms and procedures; practicing regimented cookbook applications; and solving routine problems with one-answer solutions. Simply emphasizing calculations and mechanical procedures without understanding and manipulating the deeper structures of thinking, are clearly not sufficient for competence.

Many other sources of evidence of the need for thinking and problem solving in mathematics education can be cited. One is the case of Andrey Toom, a Russian mathematician, who came to the United States expecting American students to be strongly oriented toward independent critical thinking. To his dismay, he found only a very small number of students who had experienced real thinking and problem solving. He stated, "Never before had I seen so many young people in one place who were reluctant to meet challenges and solve original problems" (Toom, 1993, p.12). Most recently, Devlin (2017) reported that due to the rapid expansion of digital capabilities brought on by reliable, high-speed desktop and cloud computing, the general population no longer needs computational skills. The essential need now is deep understanding of the procedures and the underlying concepts used by digitally implemented tools to know when and how to use these tools. He further emphasized that the skill of mathematical thinking is needed for those preparing to pursue STEM education and careers.

Recent reports comparing mathematics achievement of US students with their peers around the world show that the United States is trailing many other advanced industrial nations. Most recent results obtained in 2015 by Trends in International Mathematics and Science Study (TIMSS), which tests students in grades four and eight, reveal that ten countries (out of forty-eight) had a statistically higher fourth-grade average math test score than the United States, while seven countries out of thirty-seven had statistically higher average 8th-grade math scores than the United States (DeSilver, 2017)

Recent results for 15-year-olds on the Programme for International Student Assessment (PISA) ranked the United States 38th out of seventy-one countries in mathematics. The U.S. average was 470, below the international average of 490, with an average score range of 564 in Singapore to 328 in the Dominican Republic. Even the international average as well as the highest average of the range are far from adequate on a test based on a 1000-point scale. Andreas Schleicher, director of education and skills at Organization for Economic Cooperation and Development (OECD which coordinates PISA) pointed out the following three qualities for high performance: rigor, focus, and coherence. Schleicher stated, "Students are often good at answering the first layer of a problem in the United States.... But as soon as students have to go deeper and answer the more complex part of a problem, they have difficulties" (Kerr, 2016).

It is along this line where the RMT paradigm brings about a radical shift in the mode of mathematical learning. Through RMT, learners are deliberately and explicitly taught to think and how to apply their thinking to analyze and perform mathematical activity. Throughout the teaching and learning activity, learners articulate their thinking orally and in writing as seminal and expanding aspects of the ongoing learning process.

Nature of the Mathematics Culture

The very nature of mathematics, which originates from its culture, demands a quality of thinking for its comprehension and application. The essential components of any culture are its values, language, history, tools, and ways of doing things. RMT specifies six core mathematics concepts that collectively define the nature of mathematics and capture the values of its culture. These six core mathematics concepts are **Quantity, Relationships, Representations, Abstraction/Generalization, Logic/Proof,** and **Precision.**

Quantity is the amount or magnitude of a critical attribute or property of an object or an event, and it is determined from the action of measuring. A relationship is the connection or interaction between objects, events, or quantities. Forming, validating, and applying quantitative relationships is a central feature of mathematical activity. Thus, it is evident that mathematical tools are extensively required in mathematics for measuring and expressing quantities and for measuring, forming, and expressing quantitative relationships. Everything mathematical directly or indirectly addresses quantity. Thus, quantity and relationships are key features of the mathematics culture.

Comprehending and expressing quantities and quantitative relationships of the highly abstract concepts of time, space, motion, and energy demand the use of representations and abstract representational-relational thinking. In addition, mathematical ideas are never bound by the specifics of a single object or event, but seek to produce knowledge at a high level of generalization that speaks to the properties of behaviors of broad classes or categories of objects or events. Thus, abstraction and generalization are defining attributes of the mathematics culture.

Mathematics relies on logic, which deals with the validity of the connections among statements and the consensus of members of the mathematics community. Each foundational subject in mathematics commences with its own axioms (accepted, unproved statements) and primitive (accepted, undefined terms) from which theorems (proved statements) are constructed. Justifications and proofs in mathematics are derived through this axiomatic system and the collaborative acceptance of the mathematics community. In addition, there is an imperative need for precision or exactness in language, measurements, and computations. Thus, logic/proof and precision are innate attributes of mathematics and help to define the mathematics culture.

Mathematical Language and Tools of the Mathematics Culture

Mathematical language is both the medium and process for performing and expressing mathematical activity. Mathematical activity consists of tasks and actions that concentrate on generalizations and abstractions about patterns and relationships through chains of logical thought. Such activity can include any or all of the following:

- investigating and analyzing patterns and relationships;
- generating and describing generalizations and abstractions; and
- applying and creating new insights from generalizations and abstractions.

Mathematical language clearly and coherently expresses these tasks and actions through the integrated use of symbols, formulae, mathematical tools, and verbal definitions along with rules and propositional logic (see Kinard and Kozulin, 2008 pp. 116-119).

The tools of the mathematics culture are mathematically specific psychological tools. Mathematically specific psychological tools organize, integrate, and orchestrate learners' thinking and conceptual elements to help them construct and apply mathematical generalizations and abstractions. The prevailing culture of current mathematics instruction follows the teach-practice-test mode. This instructional model presents learners with ready-made mathematics concepts, followed by algorithmic deductive demonstrations, examples, and straight-forward problems that require direct application. On the other hand, professional mathematicians utilize existent symbolic tools or create new ones to represent, manipulate, and validate mathematical ideas. Thus, they are focused on the "process" of mathematical reasoning through the dynamic use the tools necessary. In contrast, the prevailing culture of mathematics education has learners starting with the "products" of mathematical investigation, instead of its process. This practice has learners focusing on a "mechanical" path that does not transform learners' reasoning nor provide conceptual understanding. The external forms of what could be mathematically specific psychological tools are simply representations. They are perceived by learners as pieces of information or content rather than as "tools" or "instruments" to be used to organize and construct mathematical knowledge and understanding (see Kinard and Kozulin, 2008 pp. 107-116).

On the other hand, the RMT paradigm engages each learner in seizing, as his or her own, the symbolic structure of each tool and internalizing it as a mathematically specific psychological tool. The term "symbolic tool" begins to move the representation from a static and passive form of presenting something by pointing to an active instrumental function that shapes the learner's mathematical reasoning. Consider the representation of a number line below.

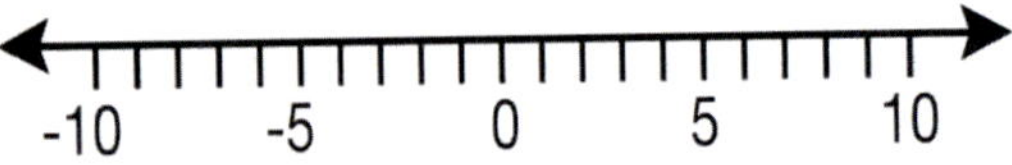

This symbolic artifact's structure is composed of a portion of a line (a quantity of linear space) that has been analyzed into equal-sized parts or segments. Each segment is a spatial interval between two successive quantities and represents the same quantitative value as each other segment. Each interval can be further analyzed into smaller equal-sized parts to increase the precision capability of the tool.

When this symbolic structure is fully appropriated and internalized by the learner as a mathematically specific psychological tool, the learner acquires and internalizes the relational aspects of its components. Next, the learner is presented relevant data and forms sets of relationships within the data, which when operated on, leads to mathematical conceptual understanding. In the case of the number line, the learner uses its internalized structure to perform the following functions:

- quantifying linear space and spatial relationships;
- analyzing, comparing, and size-ordering quantities;
- forming quantitative relationships between numbers;
- providing mathematical logical evidence regarding quantities and quantitative relationships; and,
- serving as a one-dimensional coordinate to graph solutions to linear inequalities and linear equations in one variable.

Similarly, learners engage in a series of RMT sessions devoted exclusively to the appropriation and internalization of the unique symbolic structure of each mathematical representation below. When this takes place, the internalized relational aspects of each of the following symbolic artifacts becomes a mathematically specific psychological tool: a) the base-10 number system; b) the base-2 number system; c) the base-5 number system; d) table; and, e) the two-dimensional Cartesian Coordinate System.

Mathematical Knowledge

A major goal of every culture is to transmit itself from generation to generation. The mathematics culture is perpetuated through the development and application of mathematical knowledge, which exhibits and transmits the distinguishing

Number Systems with Their Intrinsic Place Values

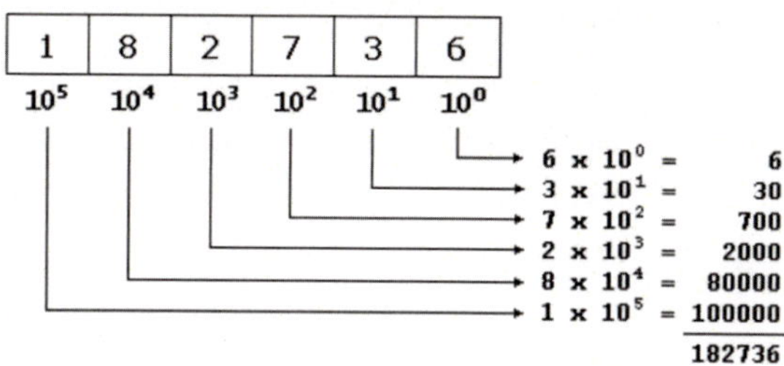

Binary Place Value Chart: Values 0-1

2^7	2^6	2^5	2^4	2^3	2^2	2^1	2^0
128	64	32	16	8	4	2	1
0	1	0	0	0	0	0	1

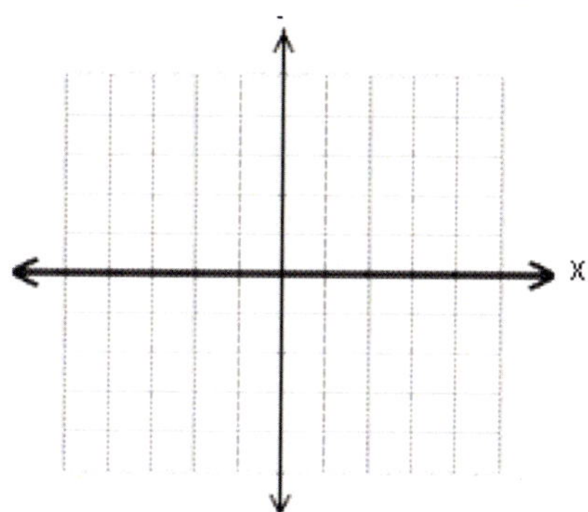

Place values of 423 (base 5)

Place value:	25	5	1
Digit:	4	2	3

Table with Columns and Rows

Heading A	Heading B	Heading C	Heading D	Heading E

X-Y Coordinate Plane

features of the culture. Mathematical knowledge exists at three levels: mathematical procedures and operations, mathematical concepts, and mathematical insights. In terms of its culture, mathematical operations and procedures provide the blue print for the way things are done within the mathematics community.

Mathematical operations and procedures organize and manipulate mathematical information in precise and meaningful ways that provide and maintain logic and coherency within the mathematics culture. In the RMT paradigm, specific, well-defined cognitive processes and established rules drive mathematical operations and procedures. In addition, as cognitive processes promote operations and procedures, they also contribute to the formation of mathematics concepts and their comprehension. For example, later in this book, we will engage you to derive some cognitive processes that are specific to the mathematics domain and guide mathematical operations and procedures, while catalyzing mathematical conceptual understanding. Some examples of these cognitive processes are quantifying space and spatial relationships; quantifying time and temporal relationships; projecting and restructuring relationships; forming proportional quantitative relationships, mathematical inductive-deductive thinking; elaborat-

ing mathematical activity through cognitive categories, and so on.

The second level of mathematical knowledge is mathematics concepts. As discussed earlier, unlike concepts in other academic disciplines, all mathematical concepts are theoretical and systemic. Mathematical conceptual knowledge perpetuates the DNA of mathematics and its culture. The RMT process engages learners to build the concept of an object or a process theoretically by constructing its ideal form and being able to experiment with it. This building means that the learner is an active agency in this construction, which produces deep understanding of the concept within the learner. Later in this book, we will model how we engage learners in building the theoretical concept of proportional quantitative relationships through the abstract general notion of parts-whole relationships and our core mathematics concepts of quantity, relationships, and logic/proof. The concept is theoretical because it does not depend on empirical data or observations for its development. This building, which is a process of cognitive conceptual construction, takes place through theoretical thinking.

Learners can mentally experiment with their theoretically derived concept of proportional quantitative relationships in the basic mathematics statement $5 = 2 + 3$, where 5 is the whole and 2 and

3 are the parts. The investigation may begin with these questions: How many portions of 3 are in 5? How many portions of 2 are in 5? From the investigation, the learner concludes that there are one and 2/3 portions of the quantity 3 in the quantity 5 and that there are two and ½ portions of the quantity 2 in the quantity 5. The mental experimentation may continue with the results from hypothetical thinking: If I double the quantity 5, that should double the portions of 2 and 3 in the new quantity. Thus, there should be four and 2/2 portions of the quantity 2 in 10 and two and 4/3 of the quantity 3 in 10. When the mental experimenting considers the whole to be one-half of 5, then the new whole, 2 ½, should contain one-half and 1/3 portions of the quantity 3 (one-half of 3 is 1 ½, and 1/3 of 3 is 1). In addition, there are one and ¼ portions of the quantity 2 in the quantity 2 ½. Learners may now formulate the folowing axiom: The addition of two real quantities to form a third quantity generates a system of proportional quantitative relationships among rational quantities. Mathematical proof of this axiom will transform it into a mathematical theorem. The theoretical cognitive-conceptual structure of forming proportional quantitative relationships is also the over-arching mathematical idea for the concepts of fractions, ratios, rates and slope in linear functions and equations. This illustrates the systemic nature of mathematical conceptual knowledge.

Mathematical insight is the highest level of mathematical knowledge. It is the apprehension of the inner nature of mathematics. Mathematical insight cannot be taught; it must be created and discovered by and within the learner. RMT mathematical learning activity aims at stimulating mathematical insight by creating a structural change in a learner's understanding of mathematical knowledge. A mere accumulation of information or skills does not qualify as a structural change. By structural change we mean a qualitative change in the learner's level of mathematical comprehension. Such a change is characterized by systemic organization, self-regulation, and transformation. A new level of comprehension should have a systemic unity so that all the elements become involved in it. In addition, the change should, on the one hand, be strong enough to withstand the temptation to solve novel tasks by reverting to previous, less advanced levels of reasoning and, on the other hand, open enough for further transformation. This new level of understanding is produced by the student's use of cognitive functions and psychological tools to organize and form relationships between supporting conceptual elements (see Kinard and Kozulin 2008, pp. 2-3).

By continuously engaging learners in constructing deep understanding of mathematics concepts, the RMT paradigm engineers a cognitive and conceptual landscape that catalyzes the creation of mathematical insight and structural changes within learners. The systemic nature of mathematics concepts synthesizes a network of dynamic inter-relationships that enlarges one's mathematical comprehension and enhances his or her mathematical ability. When a learner constructs deep conceptual understanding of a big mathematical idea, he or she establishes a new avenue through which to integrate previous mathematical knowledge into a larger, more coherent body of meaning and builds a mechanism for generating new insights or mathematical awakenings.

In summary, RMT brings about a paradigm shift in mathematics education by engaging learners in the following:

- explicitly developing thinking during the mathematics learning process;
- acquiring the mathematics culture by:
  ~appropriating and internalizing mathematically specific psychological tools based on their unique structure-function relationships,
  ~cultivating development of oral and written mathematical language;
- applying their thinking, mathematical language, and mathematically specific psychological tools to constructing deep conceptual understanding of big mathematical ideas; and
- Demonstrating strategic competence in defining, solving, and validating solutions to complex real-world problems in physics, chemistry, and other vital areas.

Summary of the Emergence of the RMT Paradigm

The first author experienced the need to explicitly teach thinking during an undergraduate physical chemistry course. All students entering the course had received no less than a B average in the math prerequisites (college algebra, trigonometry, differential and integral calculus, etc.). Yet they struggled miserably to apply the required mathematical conceptual knowledge to understand the laws and principles of physical chemistry. Conversations with the mathematics faculty and informal assessments led to the conclusion that these students' mathematics grades were based on their computational skills to perform various mathematical procedures, such as basic calculations, matrix operations, solving equations, differentiating analytic functions, integrating by parts, etc. However, their computational skills were far from adequate in helping them to grasp the concepts of physical chemistry.

The professor's strategy for teaching physical chemistry shifted to developing the students' mathematical thinking and conceptual understanding of each physical chemistry principle, and then directly applying such to related problem solving. Although the term "cognitive functions" was absent from the professor's vocabulary at that time, his idiosyncratic approach to teaching his students to think and strive for deep conceptual understanding proved effective. During the following twelve years, physical chemistry at the undergraduate college where this occurred was transformed into a capstone course, with most of the students who successfully completed the course going on to attend graduate school, medical school, or becoming employed in the chemical or pharmaceutical industry.

Twenty-three years later, the first feature of the rigorous mathematical thinking (RMT) approach appeared. This feature (Kinard and Falik, 1999; Kinard, 2001, and Kinard, 2006) focused on cognitive conceptual development and application in mathematics, science, and technology education. In 2005, Kinard and Kozulin examined the role of higher-order mental processes and psychological tools in RMT. In addition, they analyzed empirical data on cognitive and academic outcomes of the cognitively based program that aimed at producing conceptual change in underachieving students' comprehension of the mathematics concept of function.

The comprehensive theory and the instructional model for classroom practice of the RMT paradigm were published by Kinard and Kozulin in 2008. That publication also included research data from applications of the RMT model in various urban classrooms.

In RMT, as mentioned above, we claim that learners' difficulties with mathematical problem solving does not stem from the lack of mathematical knowledge but from the absence of cognitive functions. We also insist that learners' cognitive functions do not follow a natural maturational path of development: they must be actively constructed during the learning process. The RMT paradigm is based on two major theoretical approaches that allow such an active construction of cognitive functions: Feuerstein's theory of mediated learning experience and Vygotsky's theory of psychological tools (Kinard and Kozulin, 2005 and 2008).

Chapter Three

Describing the Cognitive Function Model in Rigorous Mathematical Thinking

Our Operational Definition of "Cognitive Function"

The authors' intent in the first section of this chapter is to engage you in developing an understanding of our operational definition of "cognitive function." We do this by applying the cognitive function analyzing. Let us begin with the word "cognitive" by considering that its basic meaning is thinking. Thinking involves the mental manipulation of stimuli and information to establish meaning. A function in everyday language is an action. Thus, a cognitive function is a thinking action.

The first component of our definition states, "A cognitive function is a unique, specific thinking action." The thinking action is unique because no other cognitive function can take its place. It is specific because it is exact or precise. Thus, in the

RMT model, a cognitive function is a distinct and fundamental unit of thought. It becomes so because of the mediational exposure and active practice of solving cognitive tasks.

In the physical world, each of the chemical elements is different from all the others (for example, gold, silver, iron, hydrogen, mercury, etc.). The physical and chemical properties of an element stem from the unique structure of its atom. The atom is the smallest physical particle that presents the essence or distinguishing attributes of that element. In the RMT model, a cognitive function exists at the basic or "atomic" level of thinking.

Continuing the formation of our operational definition: "A cognitive function is a unique, specific thinking action, whose exclusive meaning must be derived by the learner." The cognitive function's uniqueness is in its exact, exclusive meaning. This distinct meaning is constructed by the learner, and from its very origin it belongs to the learner. The cognitive function, so to speak, is birthed within the learner's mind and becomes the original intellectual property of the learner. The construction of its meaning by the learner makes the cognitive function "organic" for the learner. This is the seed through which human agency is cultivated.

Adding the last component of our definition:

A cognitive function is a unique, specific thinking action, whose exclusive meaning must be derived by the learner as he or she is mediated to perform and reflect on his or her performance of well-designed series of cognitive tasks.

Let us now focus on the term "cognitive tasks." Cognitive tasks are planned, organized activity that can serve as mental tools to assist in stimulating people to create, organize, and apply thinking strategies. The learner is mediated (guided, shaped, nurtured, given scaffolds) to perform tasks that require him or her to create and apply thinking strategies to solve challenging problems. It is from these types of mediated personal experiences that the learner discovers and builds the unique, exclusive meaning of each cognitive function.

Below we present written reflections by two learners on the nature of cognitive functions, following their engagement in performing series of cognitive tasks embedded in mathematical activity. Figure 3.1 shows the work of a fourth-grade learner.

It is significant to re-emphasize the understanding this young learner has succinctly formulated about what cognitive functions are and their significance in his learning. He expresses the central role cognitive functions are playing in his agency of learning by helping him

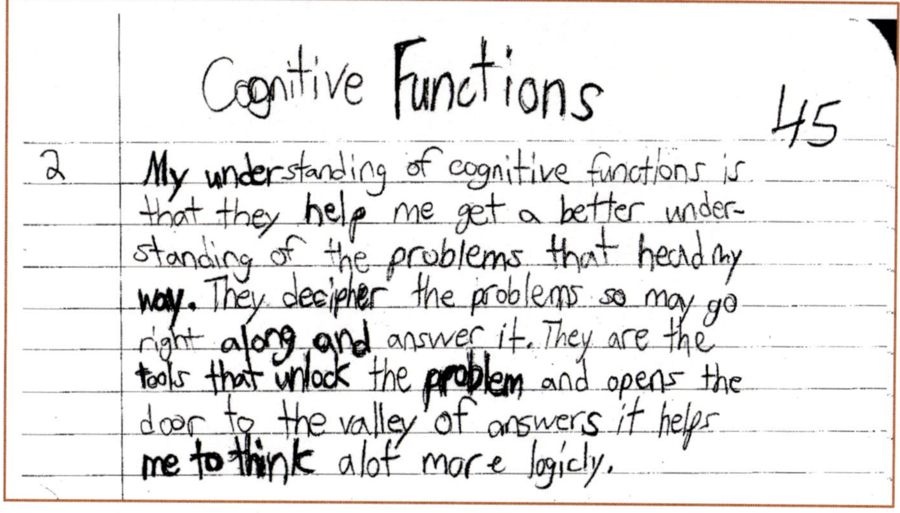

Figure 3.1: A fourth-grade learner's reflection on the significance of cogitive functions

to understand, think logically, and serve as tools to discover meaning in problem solving.

Figure 3.2 displays the work of a high school learner. She begins by referring to the atomic nature of cognitive functions and their collective role in mathematical thinking and problem solving. She emphasizes that the essence of cognitive functions stems from each function's uniqueness and exactness. Then she expresses the metacognitive nature of using cognitive functions. Cognitive functions are important in mathematics and beyond, and she states that they "govern human function in the natural world." It may be worthy to note that this learner has strong interest in vocal music and poetry. We were privately informed during RMT sessions, to our surprise, that she was being evaluated for ADHD (attention deficit hyperactivity disorder).

Roles of Cognitive Tasks and Mediated Learning Experience

Mathematical tasks, because of their very nature, have a high cognitive demand, and thus require the use of cognitive functions. Further, every learning process has a cognitive demand, a necessity for the learner to use cognitive functions to achieve the goal or goals of that learning process. In RMT, cognitive tasks in general are deliberately designed to stimulate thinking and promote cognitive function development. Thus, strategically embedding cognitive tasks in mathematical learning exercises provides the means for engaging learners in the active construction of cognitive functions during the mathematics learning process, both to enhance the learning and to promote ongoing cognitive enrichment.

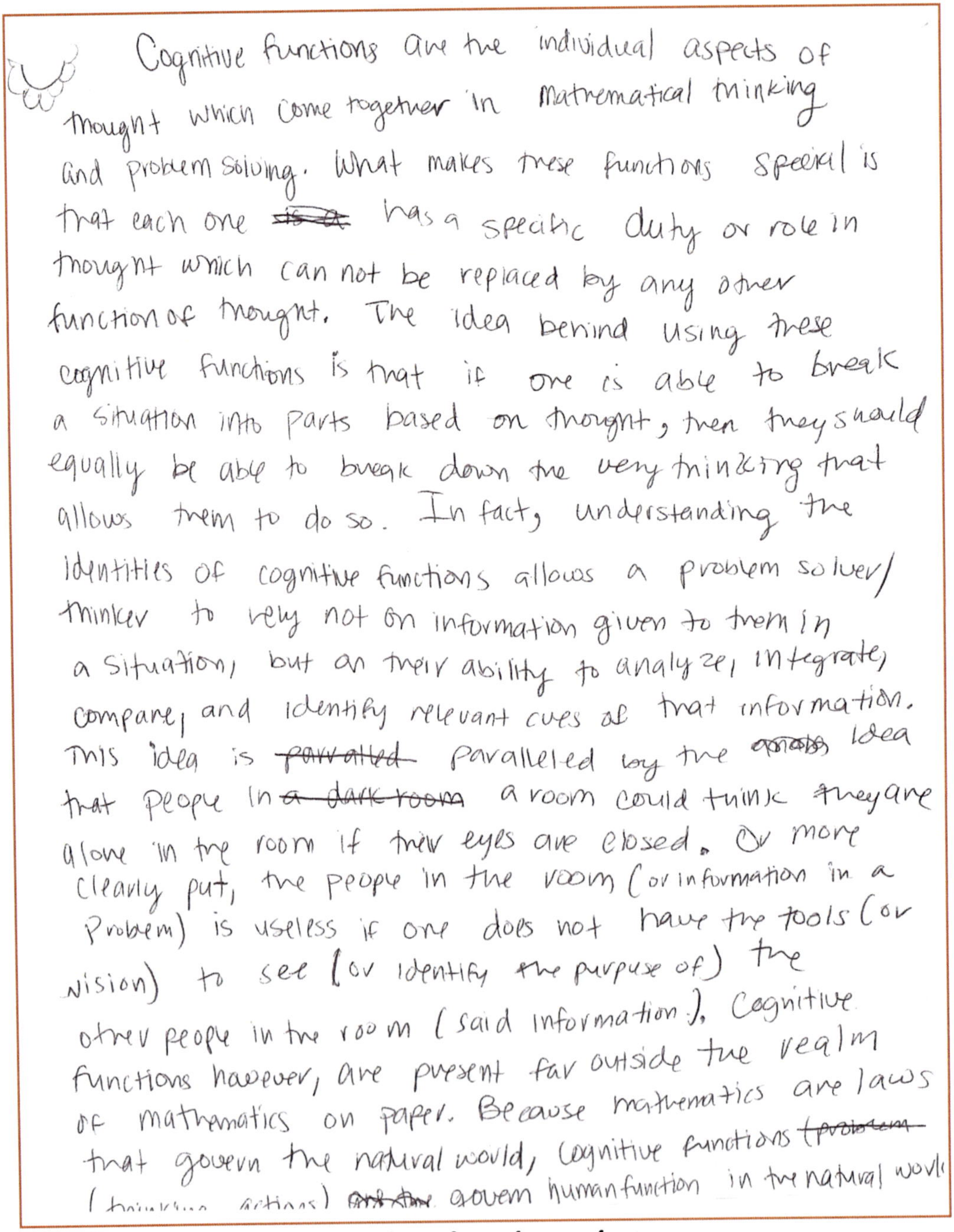

Cognitive functions are the individual aspects of thought which come together in mathematical thinking and problem solving. What makes these functions special is that each one ~~is it~~ has a specific duty or role in thought which can not be replaced by any other function of thought. The idea behind using these cognitive functions is that if one is able to break a situation into parts based on thought, then they should equally be able to break down the very thinking that allows them to do so. In fact, understanding the identities of cognitive functions allows a problem solver/ thinker to rely not on information given to them in a situation, but on their ability to analyze, integrate, compare, and identify relevant cues of that information. This idea is ~~parralled~~ paralleled by the ~~oneness~~ idea that people in ~~a dark room~~ a room could think ~~they are~~ alone in the room if their eyes are closed. Or more clearly put, the people in the room (or information in a problem) is useless if one does not have the tools (or vision) to see (or identify the purpose of) the other people in the room (said information). Cognitive functions however, are present far outside the realm of mathematics on paper. Because mathematics are laws that govern the natural world, cognitive functions ~~(problem~~ (thinking actions) ~~are the~~ govern human function in the natural world

Figure 3.2: A high school learner's reflection on the significance of cognitive functions

As discussed above, we define a cognitive function as a unique, specific thinking action whose exclusive meaning must be derived by the learner, as he or she is mediated to perform and reflect on his or her performance of well-designed series of cognitive tasks. Cognitive tasks comprise a general class of psychological tools delineated by Vygotsky, while the didactic we employ in engaging learners in performing and reflecting on their performance of cognitive tasks is Feuerstein's mediated learning experience (MLE) (Kozulin, 1998; Kinard and Kozulin, 2005, pp. 3-5; 2008, pp. 50-58).

Cognitive Tasks as General Psychological Tools

The idea of psychological tools is a central feature of L. S. Vygotsky's sociocultural theory of cognitive development (Kozulin, p. 59, 1998). His theory differentiates between two levels of mental functioning. The first level comprises elementary psychological functions, such as sensation, spontaneous or associative memory, perception, reactive attention, etc., that are naturally inherited and express our natural or innate mental abilities. The second level consists of higher-order mental functions and include focused attention, deliberate memory, logical reasoning, reading and writing skills, etc. Psychological tools, acting as sociocultural mediators, bridge the divide between elementary psychological activity and higher mental functions. Feuerstein and Falik (2010,

p. 73) refer to these tools as "vicarious" mediators.

Psychological tools are those symbolic cultural artifacts—signs, symbols, numerical systems, texts, formulae, diagrams, maps, and most fundamentally, language—that mediate our thoughts, feelings, and behaviors (Kozulin 1998, pp. 62-64). In RMT theory and practice, psychological tools comprise the vehicle for engaging learners in the active and explicit construction of cognitive functions, both during their mathematical learning and as an essential feature to effectively promote that learning. In this regard, we consider two classes of psychological tools: general psychological tools and mathematically specific psychological tools (Kinard and Kozulin 2008, pp. 52-58, 73, 95-101, 107-120).

Mediated Learning Experience: The Didactic for Engaging Learners in Performing Cognitive Tasks

The learner's rigorous engagement in performing a cognitive task takes place through the process of mediated learning experience (MLE). MLE is a quality interaction between three "actors": the mediator (an experienced person who is genuinely committed to the learner's cognitive enrichment); the mediatee, or the learner; and the cognitive task. Three essential criteria must be adhered to or applied throughout the interaction for it to qualify as a mediated learning experience. Those three criteria are **Intentionality/Reciprocity, Meaning,** and **Transcendence.**

Through MLE, the mediator must plan, deliberately guide, shape, nurture, and give scaffolds for the learner's progressive interaction with scheduled segments of the cognitive task. The mediator is intentional in constantly seeking to secure the learner's reciprocity (input and participation) by:

- modeling and displaying a state of vigilance;
- selecting, isolating, framing, and helping the learner to interpret key aspects of the task; and
- creating an atmosphere that encourages the learner to activate his or her prior knowledge, draw from everyday experiences, and use everyday concepts to connect with the cognitive task.

Through Intentionality/Reciprocity the mediator and learner together become active in posing questions, shaping/reshaping responses, debating and justifying.

Meaning is mediated at two levels: by focusing on creating insight and understanding (at the cognitive level) and by charging this interaction with excitement, enthusiasm, and motivation (at the emotional level). Mediation of meaning emphatically takes place as the learner is facilitated to extract, formulate, and construct the unique, exclusive meaning of the cognitive function from performing and reflecting on the performance of cognitive tasks, along with forging that meaning into an action word or phrase. The meaning of the cognitive function must be perceived, organized, and synthesized directly by the learner as he or she explores, manipulates, interprets, and performs the cognitive task.

Transcendence is essential for the interaction to truly qualify as a process of experience rather than a momentary encounter with stimuli or content. It involves moving from the "here and now" of doing to a broader spectrum of development. Mediating Transcendence helps the learner to cultivate the need and the ability to connect features of the interaction and the cognitive tasks to his or her ever-expanding need for inquiring, forming relationships, understanding, and reflective thinking. The apex of Transcendence is mediated when the learner spontaneously applies the meanings of a cluster of cognitive functions to achieve a mathematical academic standard.

Now, after elaborating on the roles of cognitive tasks and the MLE dynamic in operationalizing our definition of a cognitive function, we revisit that definition. We define a cognitive function as a unique, specific thinking action, whose exclusive meaning must be derived by the learner, as he or she is mediated to perform and reflect on his or her performance of well-designed series of cognitive tasks. Thus, a cognitive function that has been completely developed by the learner is a mediated internalized structure of conceptual meaning that equips the learner with the need, ability, and intrinsic motivation to perform a specific mental action on an object, a piece of information, or a field of data.

A cognitive function derived in this manner consists of three distinct but interacting components that work in a relationship with each other to provide the cognitive function with its integrity and distinct mental activity or thought process. These three components of each cognitive function are the **conceptual** component, the **motivational** component, and the **operational** component.

The Anatomy of a Cognitive Function

The Conceptual Component

The unique, specific meaning of each cognitive function that is derived by the learner during the progressive performance of cognitive tasks is conceptual in nature. For example, the cognitive function analyzing means breaking something down into its parts or decomposing something. The conceptual nature of analyzing is whole-to-parts, composition-to-segments, unit-to-subunits, etc.

Consider the cognitive function **integrating,** which means putting or merging things together to construct something different or composing something that is distinct from the parts that are involved in building it. The conceptual component of integrating is parts-to-whole, segments-to-unity, components-to-composition, individual items-to-category; elements-to-set. Another example is the cognitive function **providing logical evidence.** When the learner is providing logical evidence, he or she is giving clues, hints, supporting details and proof that make sense in validating (verifying) or denying a hypothesis, a claim, or a statement. The conceptual component of this mental activity is educated guess/justification, position/foundation, inference/conclusion.

The conceptual component of the cognitive function provides a "steering" mechanism to the mental activity. This component can be further viewed as an interaction between procedure and purpose—the mechanism of the function as it is guided and shaped through a conceptual meaning that becomes a conceptual action. While the nature of the stimuli will influence the details of how the function will be executed, the action is driven by the same conceptual meaning, whether the object is a physical entity, mathematical problem, play, movie or written text.

The Motivational Component

The motivational component of a cognitive function is an internal feature that activates, mobilizes, and energizes the function to act on the environment with a coordinated unit of conceptual meaning. This feature is derived from two interdependent factors—a function-bound needs system and the learner's growing perception of the importance or benefits of carrying out the function.

A cognitive function, like any other voluntary motion or action, only takes place under a need—a stimulant that becomes a cause or catalyst for producing a required effect or outcome. The function-bound needs system commences to be established when the learner: 1) begins to develop the function through direct participation in specially designed learning activity; 2) discovers a problem or difficulty that directly requires the use of the cognitive function; 3) applies the function to help address the problem or difficulty; 4) becomes aware of specifically how use of the function helps to meet that specific need; and 5) expands this repertoire by applying the cognitive function to solving other problems in different contexts.

The need for the cognitive function **focusing** appears in the forms of a lack of clarity and insufficient details when the learner is confronted with a new or complex situation that is difficult to approach. After investing appropriate time and attention to using past experience to analyze the new situation and carefully explore each part, the learner experiences success in beginning to understand what is taking place. Receiving an internal reward for investing time and energy to concentrate and pay attention to specific parts of a previously vague or unknown situation begins imbuing the thinking action of focusing with a motivating feature that is specific to its dynamic meaning.

A function-bound needs system is the crystalized, flexible, and fluid habit of executing this cognitive function in a wide range of situations with different levels of complexity, novelty, and abstraction. The automatic behavior embodied in this needs system is intrinsic to the function because it lies within the internal mechanism of this thinking action. This means that the activation of this cognitive function is detached from sources in the external environment or the extrinsic need that initially produced it (See Feuerstein, Rand, Hoffman, and Miller, 1980 pp. 116 and 273; Rand, 1991, pp. 80 and 81).

As stated earlier, the second factor that produces the motivational component of a cognitive function is the learner's awareness of the perceived importance and benefits of executing the function. This factor is developed as the learner is guided to practice the use of the conceptual know-how of the cognitive function.

This practice is deliberately designed to take place through various modalities and across a range of problem-solving contexts, while the learner is clearly and explicitly becoming fully aware of the benefits derived from the cognitive applications and of his or her academic competence demonstrated from performing the tasks and solving the problems. After this function-produced competence has been displayed, the learner is guided to acquire a feeling of competence in his or her developing capacity to perform the specific cognitive function (See Skuy, 1996, pp. 27-32). The acquisition of a feeling of competence helps the learner to gain the self-confidence needed to continue to engage successfully in performing the cognitive function.

It should be obvious that the two factors that lead to the development of the motivational component of a cognitive function are in both an interdependent and dialectical relationship with each other. As the function-bound needs system becomes more and more crystallized and flexible, the learner's awareness of the benefits of applying the cognitive function becomes more pronounced and visible, thus enhancing the learner's feeling of competence and increasing his or her willingness to invest in the successful practice of the function, making it more robust. As the needs system moves toward greater internalization and flexibility, the greater is the capacity and efficiency of the function. This reduces the effort the learner needs to invest to activate and mobilize the function. The habit-forming nature of the needs system is informed by an induced anticipation of success in the future execution of the cognitive function, incorporating and automatizing this anticipation into the structure of the needs system.

The Operational Component

The operational component of a cognitive function brings about the organization, integration, and coordination of the other two components—the conceptual and motivational components—into a unit or quantum of discrete energized meaning that operates on targeted stimuli, information, or process within the experience of the learner. This component is a synergized interaction that is informed by and impregnated with meaning from the conceptual component while being activated. Cognitive functions never exist in isolation. Through the operational component, one or more clusters of cognitive functions act upon objects and data to produce within the learner a fabric of conceptual knowledge and understanding.

In summary, the RMT model is based on two major theoretical approaches that allow such active construction of cognitive functions: Vygotsky's theory of psychological tools and Feuerstein's concept of mediated learning experience. The RMT paradigm defines a cognitive function as a unique, specific thinking action whose exclusive meaning must be derived by the learner, as he or she is mediated to perform and reflect on his or her performance of a well-designed series of cognitive tasks. The mechanism that leads to the active construction of cognitive functions during the learning process is inherent within this operational definition of a cognitive function. Thus, a cognitive function that has been completely developed by the learner is a mediated internalized structure of conceptual meaning that equips the learner with the need, ability, and intrinsic motivation to perform a specific mental action on an object, a unit of information, or a field of data.

Feuerstein's Instrumental Enrichment Program as a Rich Source of General Psychological Tools

Feuerstein's Instrumental Enrichment (FIE) cognitive intervention program offers one of the richest sources of general psychological tools (Feuerstein et al., 1980, pp. 115-256; Feuerstein et al., 2006, pp. 211-309; Kinard and Kozulin 2008, pp. 90-101). FIE is one of the main applications of Feuerstein's theory of MLE to increase human modifiability, making learners more amenable to direct learning situations. The FIE program consists of 14 systematically organized learning activities to build or strengthen cognitive functions, with active mental and procedural learning, using discussion, problem solving, and a "paper and pencil" response modality. Cognitive functions, approached in this way, enable the students to become amenable to restoration or acquisition of cognitive functions through a systematic and structured mediated learning experience.

Feuerstein's concept of cognitive functions originates with and is centered around "deficient cognitive functions." This genesis was seeded during Feuerstein's groundbreaking work with children and youth who were survivors of the Holocaust and were gathered in camps in Southern France and Morocco. Psychologists and social workers reported that these children and youth were very low functioning and perhaps mentally retarded. Utilizing a multitude of diverse clinical observations, Feuerstein generated a list of "deficient cognitive functions" that frequently lead to inadequate problem solving.

In the RMT paradigm, we strategically select cognitive tasks from instruments in FIE as general psychological tools to build cognitive functions based on the RMT model. As you can see, the RMT operational definition of cognitive function is distinctively different from Feuerstein's construct. Thus, the RMT model neither promotes the practice of FIE, nor is it a result of bridging from FIE mediation. ■

Chapter Four

Engaging Learners in Building Cognitive Functions in the Mathematics Learning Process

Justifying and Validating Mathematical Outcomes

Focusing, Selecting Relevant Cues, and Providing Logical Evidence

The following vignette and discussions (Exhibit #1) demonstrate how a teacher uses MLE to engage learners in performing cognitive tasks to build cognitive functions.

Exhibit #1

<u>Vignette</u>

Mediator: What is this person doing?
Learners: Staring at the paper.
❑Reading.
❑Trying to figure something out.
❑Maybe he is studying.
Mediator: When you are reading, studying, or trying to figure something out, where is this action taking place?
Learners: In my head.
❑In my mind.
❑In my brain.
Mediator: If he is doing something in his mind, what is he doing?
Learners: Thinking!
Mediator: Great! Let us write the title of the first section in our Journal of Reflection. The title will be made up of two words. The first word will mean what the person is doing in the picture. What did we agree the first word will mean?
Learners: Thinking!
Mediator: That's correct. I will spell it as you write it down. C-o-g-n-i-t-i-v-e. Let us pronounce this word. What does it mean? [The mediator

is deliberate in insisting that all learners are pronouncing this word correctly and beginning to internalize its meaning]. The second word is spelled F-u-n-c-t-i-o-n-s. Pronounce it.

Mediator: What is a function in everyday life?

Learners: The way something works.
❑The use or purpose of something.
❑A process.
❑How something moves.

Mediator: Wonderful. Thank you for your thoughtful responses. Would the word "action" be a simple way of saying it?

Learners: Yes!
❑So cognitive functions are thinking actions!

Mediator: Very good! So you made the claim that the person is thinking . . . How do you know that the person is thinking?

Learners: He's staring at the paper.
❑Yeah, he is zeroing in on something specific on the paper.
❑He's on point!

Mediator: Okay. When you are zeroing in on something or staring at something specific or on point, what are you really doing?

Learners: When I do something like that, I am focusing on something specific.
❑That's like being on point … concentrating.
❑When I do that I'm blocking out everything else except what I'm paying attention to.

Mediator: Wow! It is amazing how all of you together constructed the name of your first cognitive function and derived its meaning. What is a cognitive function?

Learners: A thinking action! (in concert)

Mediator: You said when you are staring at something or zeroing in on something specific and ignoring everything else, like the person in the picture, you are focusing. **Focusing** is the name of your first cognitive function. Write it down in your journal. [Pause] When you are focusing, what exact action are you taking in your mind? I knew that you really got it when you used phrases like "on point" and "blocking out everything else."

Learners: We are paying attention.
❑Yeah, we are concentrating.

Mediator: Wow! You have just constructed the name and meaning of your first cognitive function or thinking action. Let's summarize. The

> name of the cognitive function is focusing. What are you doing in your mind when you are focusing?

Learners: We are concentrating on something.
❑Yeah, we are paying attention.
❑Putting our undivided effort on something specific.

Mediator: I need each of you to create an illustration, which is a drawing or a diagram, that captures your exact meaning of the cognitive function focusing. Make your illustration so impactful that anyone who sees it will clearly get from it the meaning of Focusing. (Three learners' illustrations are shown in Figure 4.1 below.)

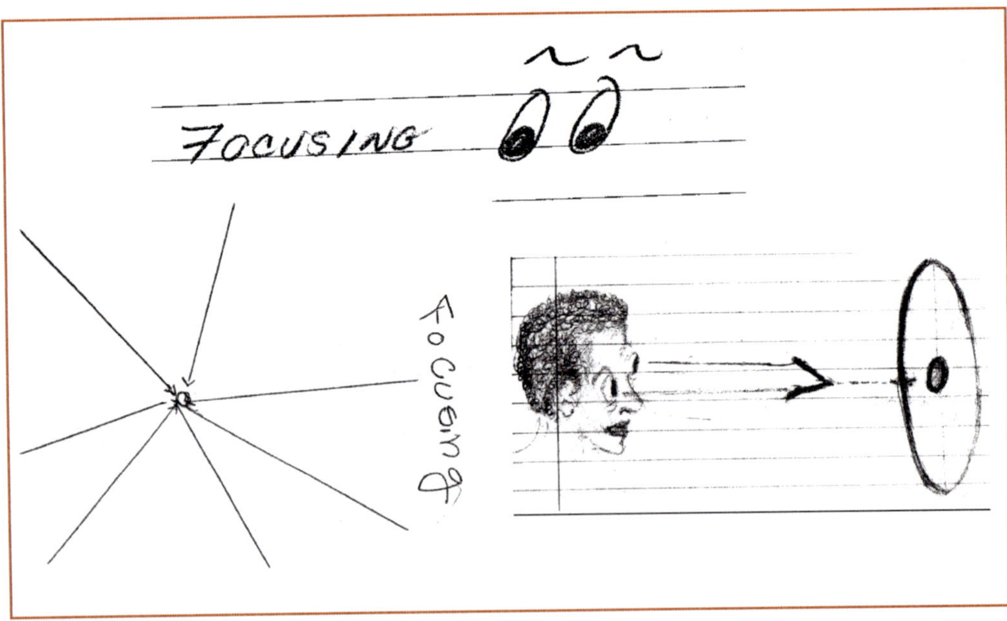

Figure 4.1.: Schemas of learners' meaning of the cognitive function focusing

Discussion

Notice the mediator's intentionality in both activating the learners' prior knowledge and guiding them to utilize their everyday language and concepts to actively formulate and capture the meaning of the cognitive function Focusing. This begins a process of channeling the learners into the progressive pathway of internalizing, owning, and making flexible and robust this distinct mental activity.

<u>Continuation of Vignette</u>

Mediator: What are you doing when you make a claim in science?

Learners: You are giving your point of view.

❑Maybe, but I think you are planting your ground work on an idea.

❑It's like making an assumption.

❑I see it like making an assumption based what you observed.

Mediator: I like what I'm hearing. Have you heard of the word proposition?

Learners: Yes. Is that what you would call an educated guess?

❑That's a hypothesis.

Mediator: Very good! When you made the claim the person was thinking, were you giving a hypothesis?

Learners: Yes!

Mediator: How do you know the person is thinking?

Learners: He is looking very intense.

❑His posture . . . He is leaned back in the chair.

❑One hand is on his chin, and a paper and pencil are in the other hand.

❑Yeah. There are books on the floor beside him. He's thinking all right!

Mediator: Wonderful. You formed the hypothesis that the person is thinking. I asked you how did you know and you started giving me things—his posture, his facial expression, what is in his hand, books on the floor beside him. What are these things you are giving me?

Learners: They are clues.

❑Pieces of support.

❑Yes, these are supporting details.

Mediator: But what are these details supporting?

Learners: Oh, our hypothesis.

❑So we were giving you evidence.

Mediator: Great! What is evidence?

Learners: Facts.

❑More like proof.

❑Backup!

Mediator: Did this evidence make sense?

Learner: Yes. It connected with the situation.

Mediator: That means it was logical. What is another word that means giving?

Learners: Providing.

Mediator: What were you providing to support your hypothesis?

Learners: Evidence.

❑Logical evidence.

Mediator: You just constructed your second cognitive function, **providing logical evidence.** Look what you have accomplished. You formed a hypothesis that the person was thinking. I asked, "How did you know the person was thinking and you immediately started giving me clues, hints, supporting details, and proof to let me know that you know the person is thinking. Not only did you develop the name of the cognitive function, you also derived its unique and specific meaning. This is fantastic work!

(Figure 4.2 presents two learners' illustrations of the meaning of the cognitive function providing logical evidence).

Figure 4.2: Schemas of learners' meaning of the cognitive function providing logical evidence

Continuation of Vignette

Learners: This really has me thinking deeply.

❑Me too. I had to keep focusing and staying with the flow.

❑This has me digging for what I already know and connecting it with what we are now doing. This is exciting.

Mediator: A lady who has been coughing, sneezing, and feeling exhausted for several days goes to her doctor. What could be specific pieces of logical evidence the doctor will collect to determine what might be wrong with the lady? Come to the board; write each piece of logical evidence.

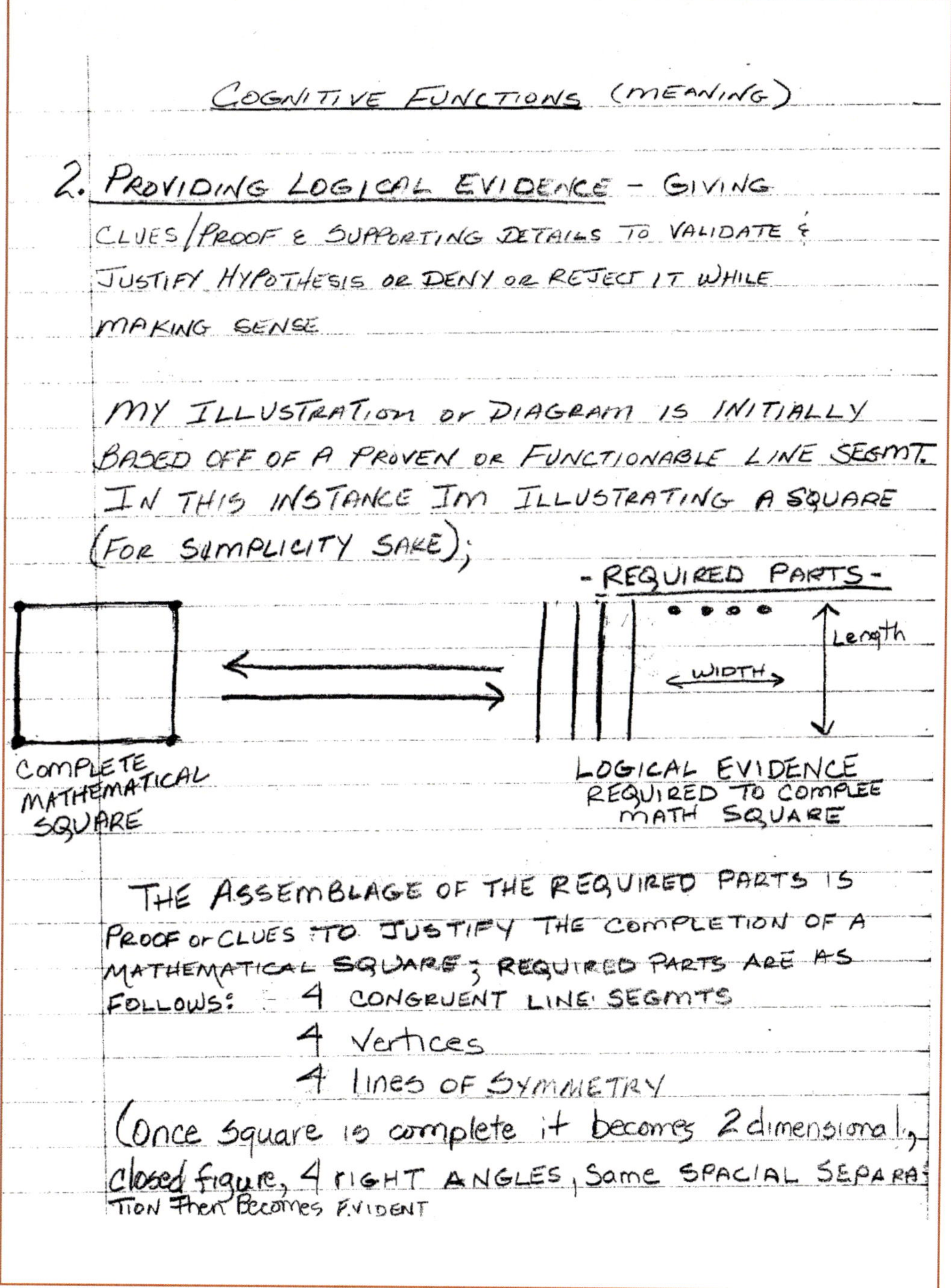

Figure 4.2: Schemas of learners' meaning of the cognitive function providing logical evidence (continued)

[Learners' responses: Her answers to questions about how she is feeling, when did this start, etc. Weigh her. Take her temperature, blood pressure, pulse, a throat culture, x-ray, etc.]

Mediator: Great job! These are pieces of logical evidence the doctor may collect. What makes the lady's weight a piece of logical evidence?

Learners: Suppose the scale shows that she weighs 135 pounds. When the doctor compares her new weight with her previous weight 6 weeks before and discovers that she has lost 20 pounds her weight is an important clue or hint that something may be wrong.

❏This is a good point. This will give the doctor a reason to ask her if she was on a diet or working out more than usual or has she been sick for a while.

❏Yeah, I see. We ask the same question about her temperature or blood pressure. If her temperature is higher than normal, it could be a clue that she has an infection.

Mediator: So what are we saying about a piece of logical evidence?

Learners: A piece of evidence may be a guide or a hint on how to search for more evidence.

❏In some cases, in order to provide logical evidence, we need more than one piece of evidence.

❏Maybe these are not really pieces of evidence; they may be hints or guides or clues.

Mediator: Now you are building a new cognitive function **selecting a relevant cue.** Is this cognitive function different from providing logical evidence?

Learners: Yes. When providing logical evidence, we have to give enough supporting details to prove or justify the hypothesis. That may take more than one piece of evidence.

❏When we are selecting a relevant cue, we are taking something as a guide or a hint that could help us accomplish something later.

Providing Mathematical Logical Evidence

a. What number belongs in the box to make the number sentence true?

$8 + 9 = 17$

$17 - 9 = $ ■

8	9	17	26
A	B	C	D

Providing additional mathematical logical evidence:
$8 + 9 = (5 + 3) + (5 + 4) = 10 + 7 = 17$

b. Solve the following problem and provide your mathematical logical evidence to support your result

$379 + 298 + 455 =?$

See a learner's work in Figure 4.3.

c. If $b = 2e$ and $r = 1.5e$, provide all of your mathematical logical evidence to validate that $r = (b \div 4) \cdot 3$

This problem is used to engage learners in mathematical discourse. Learners are assigned to work in small collaborative teams. After a period of time, the mediator has the teams to present their work and justify their results. Learners are led to compare and discuss the different approaches to solving the problem.

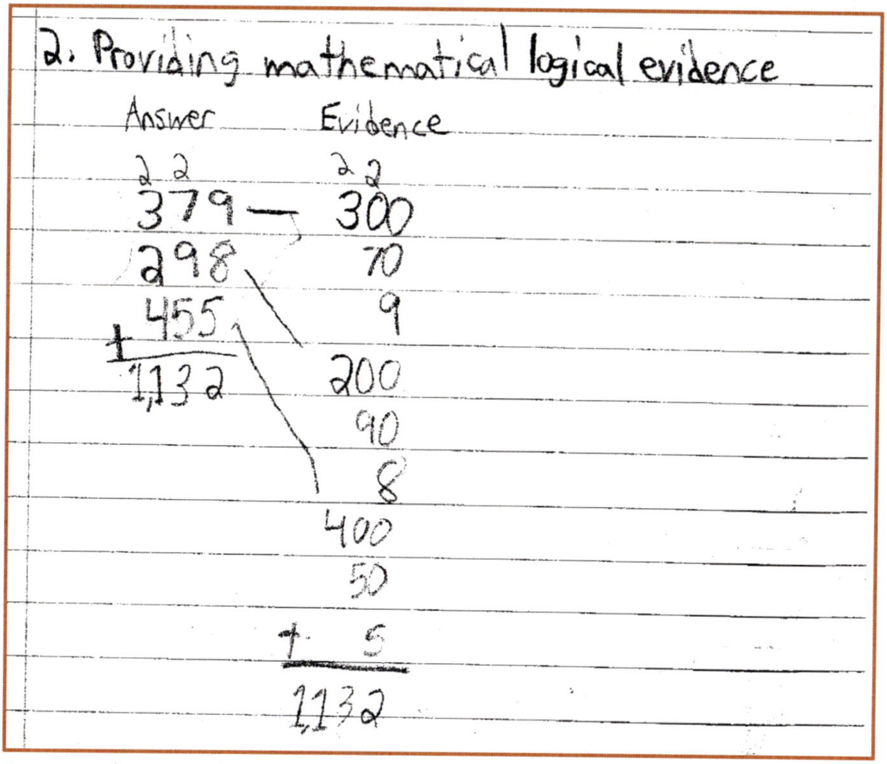

Figure 4.3: A learner's work in solving a problem and providing mathematical logical evidence

Establishing a Cognitive Structure for Conceptual Understanding

Analyzing and Integrating

When mediating learners to derive the unique meanings of the cognitive functions **analyzing** and **integrating,** we intentionally choose tasks that require moving from whole-to-parts, entire-to-segments, complete-to-portions, full-to-pieces, total-to-each, and vice versa.

Vignette

Mediator: Take your time and carefully examine this drawing (see Exhibit #2 below). [Pause] What do you see?

Learners: It seems that an oval is breaking into pieces.
❏Yeah, but it's not shattered. The pieces are exactly equal.
❏You are right. The four parts are separated and you can tell they are coming from the complete oval.
❏I get what you are saying. The parts belong to the full shape, but they are standing out like clues or hints.
❏These are relevant cues to help us understand that we are breaking the oval into parts.

Mediator: Powerful! This is great thinking. You used the cognitive function selecting relevant cues to begin building the cognitive function analyzing.

Learner: But I see something else too. As we move back into the page the parts are coming together to make the oval.

Mediator: Wow! You just derived the meaning of the cognitive function Integrating.

Discussion

Learners work together to determine that when:

- **analyzing,** they are breaking something down into its parts or they are decomposing something; and
- **integrating,** they are bringing or merging things together to make something different or they are composing something from its parts.

The cognitive tasks from FIE instrument Analytic Perception contained in Exhibit #3 create the need for learners to use the two cognitive functions, analyzing and

Exhibit #2

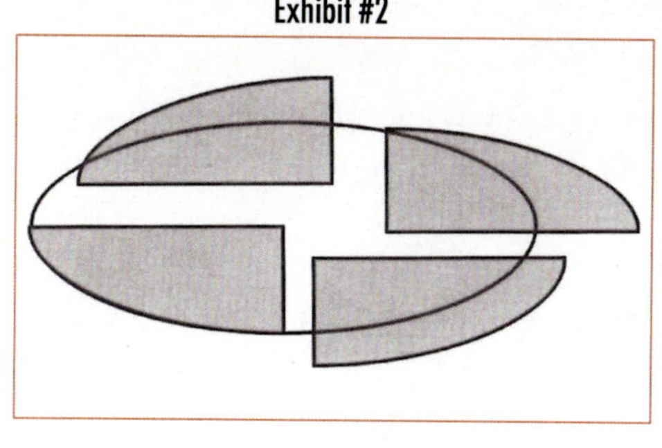

integrating together as a unit, along with the cognitive functions focusing, selecting relevant cues, and providing logical evidence. The written directions at the top of the page can be translated: "Can you integrate the parts in each frame to compose the complete or whole design?" First, one has to continually focus and analyze the complete design by decomposing it into its parts.

The first design on the left in the first row is a square. By analyzing it, one produc-

es eight parts—six congruent squares and two congruent rectangles. An inventory of the parts reveals seven congruent squares and one rectangle, all of which are of the same size as required from the design. These two actions create the relevant cue for determining the nature of the error. One strategy for correcting the error is to eliminate one of the squares and construct another rectangle that is congruent with the one given.

Exhibit #3

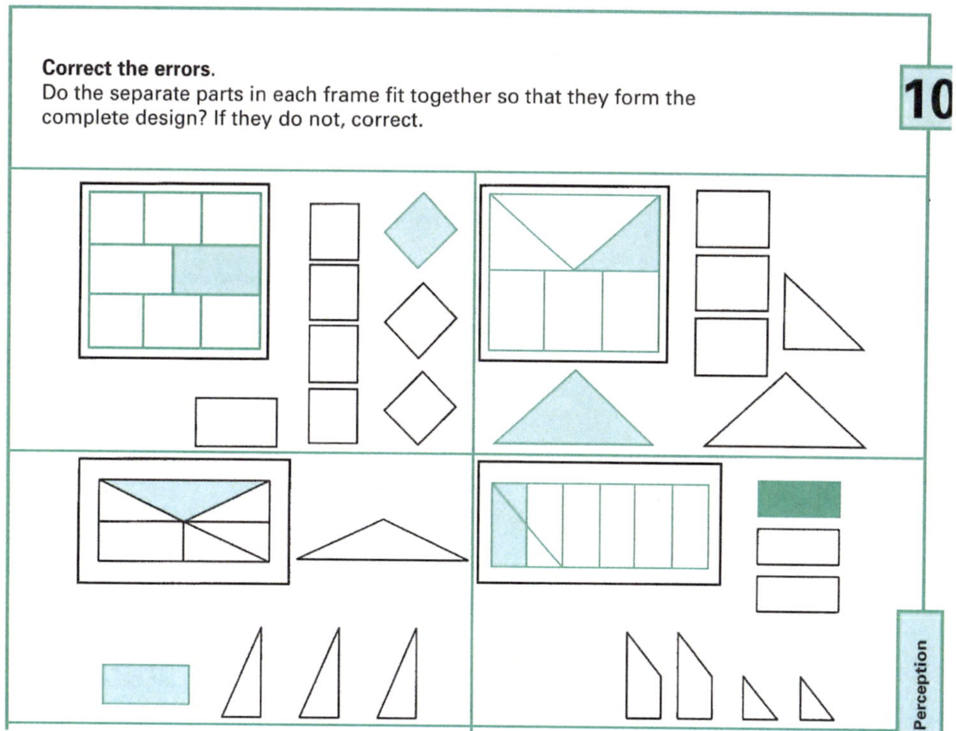

Notice the complexity and variations in the tasks, which build the learners' flexibility in dynamically analyzing-integrating as a cognitive structure, while meeting the demand for focusing, selecting relevant cues, and providing logical evidence. Here is where the mediator must be academically rigorous in intentionally and methodically guiding each learner to explicitly express how he or she is using the exclusive meaning of each cognitive function to support his or her performance of the tasks. It is essential for each learner to construct detailed written reflections on his or her work in performing these tasks.

Figures 4.4 – 4.7 present learners' work from analyzing geometrical figures that are increasing in complexity. In each case, analyzing the figure generates its critical attributes, properties, characteristics, features, or traits. This type of progressive cognitive application helps learners to discover that analyzing is an important cognitive process for constructing deep conceptual understanding.

In addition, this progressive application of analyzing leads learners to derive the exclusive meaning of the cognitive function **labeling,** giving something a name based on its critical attributes. The label "square" captures the critical attributes or properties produced in Figure 4.4. In Figure 4.5, the critical attributes: 2) "two congruent line segments"; and 4) a "90⁰ angle and two 45⁰ angles, provide the mathematical log-

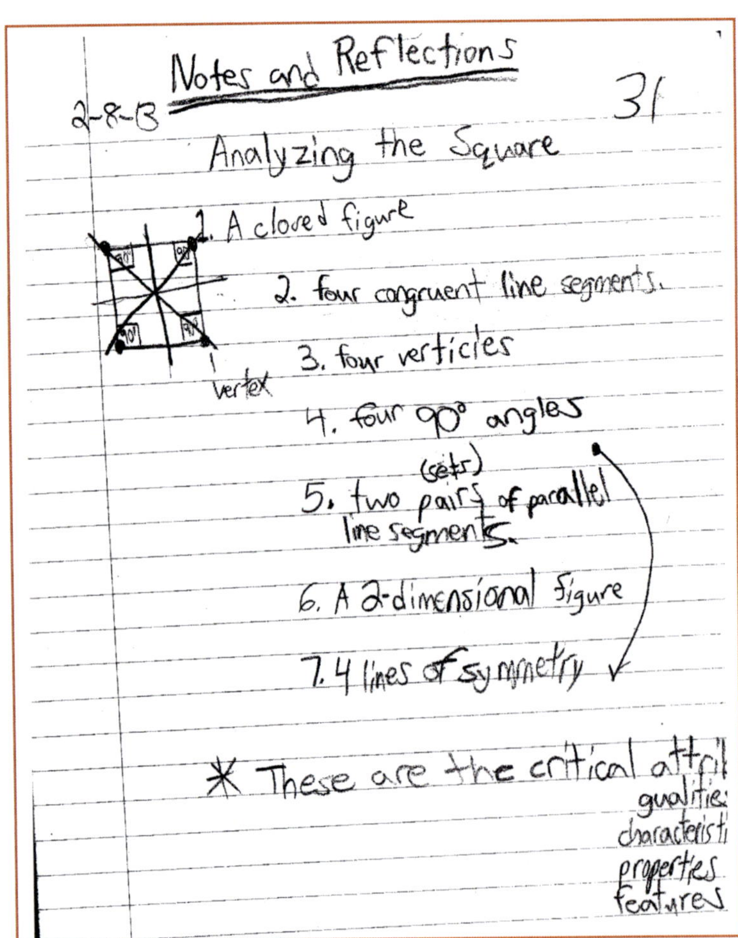

Figure 4.4: A learner's work on analyzing a square

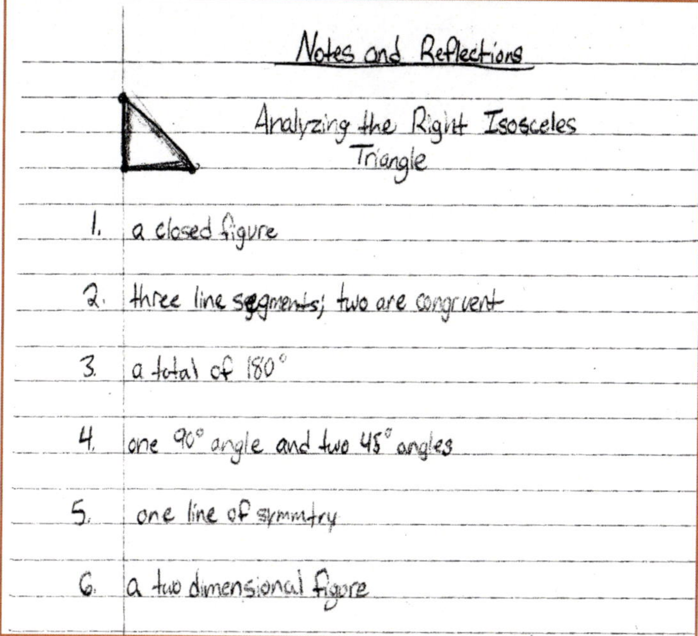

Figure 4.5: A learner's work on analyzing a right isosceles triangle

"square" based on the critical attributes or properties it possesses. The unique meaning of the cognitive function **visualizing** is demonstrated in Exhibit #5. First, a verbal concept of an object is presented on the left, followed by a visual image of that object. The first example, the verbal concept "square" in the learner's mind stimulates the learner to construct a figural or pictorial image of that concept. The learner is visualizing what he or she

ical evidence to label this figure a right isosceles triangle.

Notice in Figure 4.6 that the learner constructs the mathematics concept of collinear points.

Labeling and Visualizing

Exhibit #4 illustrates the meaning of the cognitive function labeling. Properties or critical attributes of an object are verbally presented, followed by the statement "This is a square." This mental action is giving an object the name

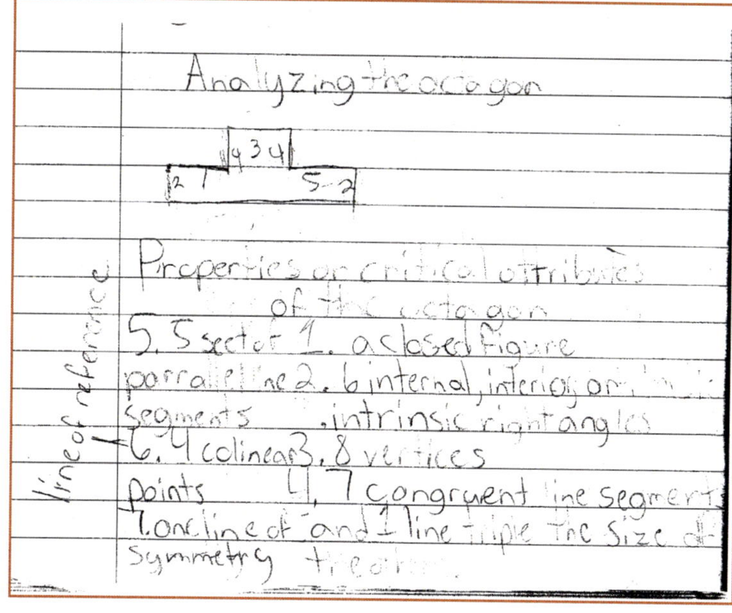

Figure 4.6: A learner's work on analyzing an octagon

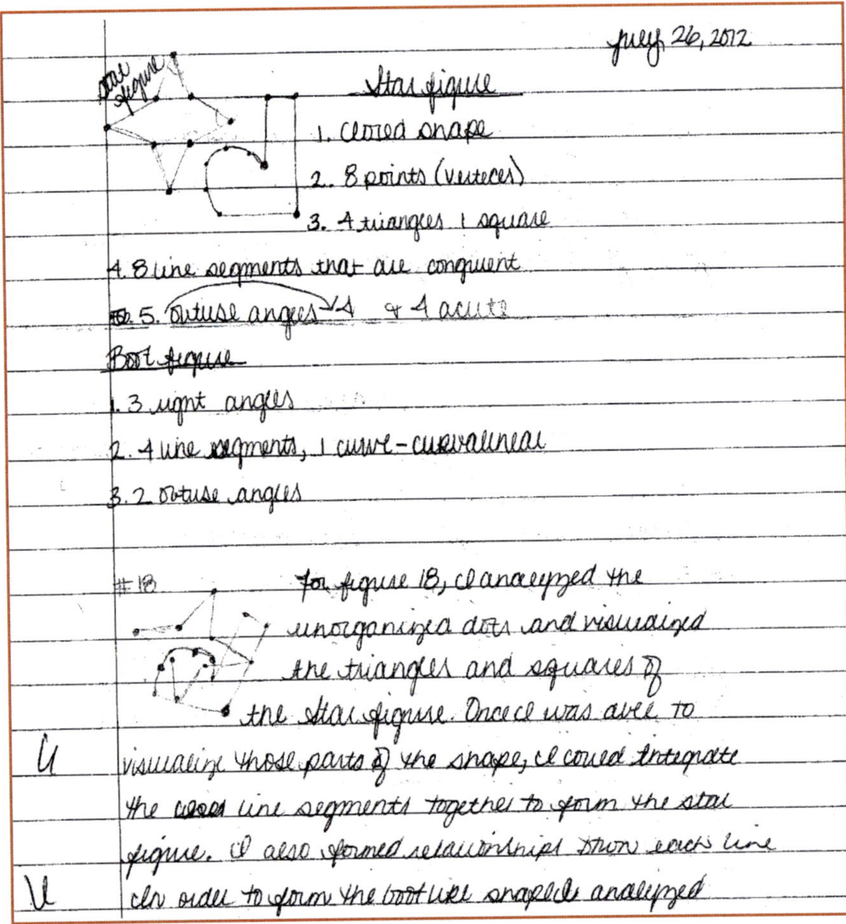

Figure 4.7: A learner's work in analyzing a star and a boot

has previously labeled, stemming from having previously analyzed the object. In the second example, when the learner experiences the verbal concept "an equation," the mental image of the symbolic expression of a mathematical equation is formed. Thus, the cognitive functions labeling and visualizing possess a natural synergy.

Developing Deep Conceptual Understanding of Quantity and Quantitative Relationships

Mediating Learners in Constructing the Core Mathematics Concept of Quantity

The RMT paradigm emphasizes that effective mathematics education requires acquisition of the mathematics culture. Key features of any culture are beliefs, values, language, tools, history, and the

Exhibit #4

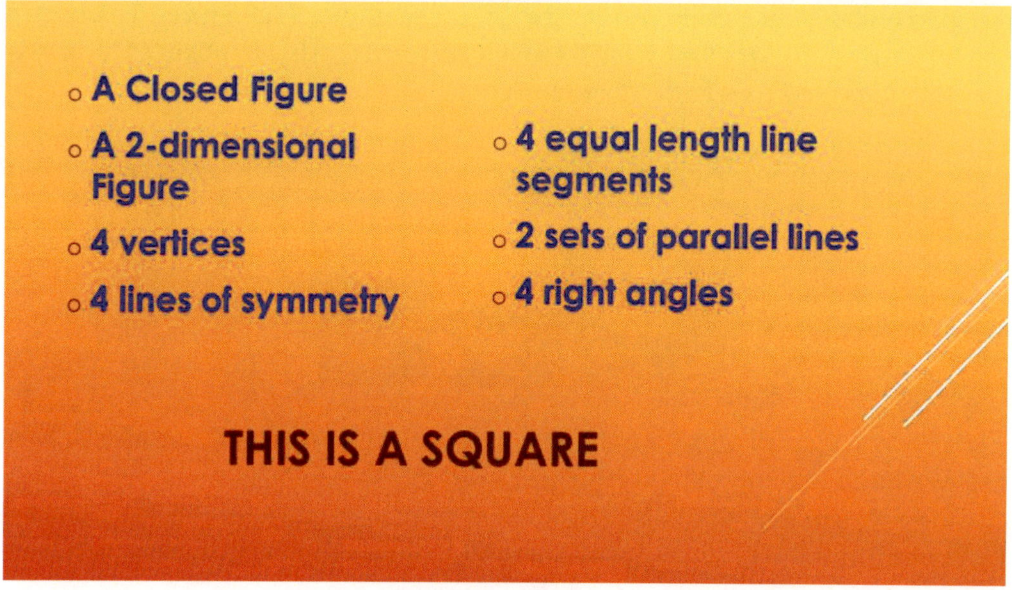

Exhibit #5

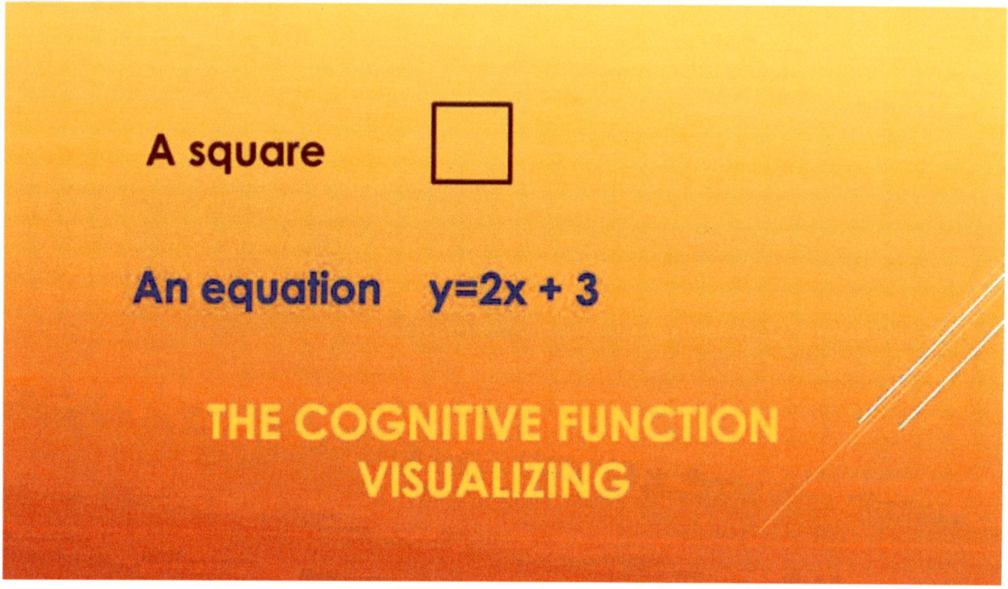

way things are done. We identify six core mathematics concepts that, together, capture and embody the nature of the mathematics culture. These six core concepts are **Quantity; Relationships, Representations, Abstraction/Generalization, Logic/Proof,** and **Precision.**

Exhibit #6

The process of mediating learners in constructing the core concept of quantity begins with presenting Exhibit #6 above.

Comparing

What is changing in the glass as we move from left to right?

How _________ of the liquid in the glass is increasing as we move from left to right?

Give four words that mean how _____________ of something.

The _______________ of something

The _______________ of something

The _______________ of something

The _______________ of something

What action do we take to determine how much of something?

What action do we take to determine how many?

Discussion

The mediator has three intentions in seeking reciprocity from the learners from the content of this exhibit:

1. To engage the learners in applying their ready-made perceptions of "Comparing" to address each of the questions posed;
2. To activate the learners' prior knowledge and everyday language and concepts to begin constructing the mathematics concept of quantity; and

3. To bring about a structural change in the learners' comprehension of mathematical knowledge.

The mediator draws learners into focusing on the picture and creates the context "How _________ of something?" to stimulate them to produce the term "much," which is impregnated with cultural meaning. The question is designed to produce the terms amount, quantity, value, and magnitude of something (in this case the liquid in the glass), which can only be determined from the action of measuring.

Further meaning and transcendence are mediated when learners are guided to present and discuss examples of quantities of objects or substances they encounter in everyday living and describe how are they measured and expressed. The presentations and discussions produce the need for discourse on the structure and function of each tool and a debate on similarities and differences of the actions of measuring and counting. The following understandings emerge from these engagements:

- every quantity has a unit of measure.
- number is a part of the tool system for measuring quantity; and
- the mathematics concept is quantity rather than number.

Comparing through Superordinate Concepts

An RMT strategy that fosters conceptual thinking and understanding is comparing through superordinate concepts. A structure through which we begin mediating learners to engage in this construction is presented in Exhibit #7. Visualize drawing an imaginary vertical line between the two objects, so that you will have an object on the left of this line and an object on the right of this line. Below these two objects is a table that has five columns and each column has a heading. The first column on the left has the heading "Concept by which to Compare." The next column is "Object on Left," the third column is "Object on Right," the fourth column is "Similarity," and the fifth column is "Difference."

The first column is already filled for the "Concept by which to Compare." During the actual mediation, we begin with this empty column, and only make an entry as we proceed in the process. The first concept is figure. Learners are mediated to build a clear understanding of the concept. Next the object on the left will be analyzed (broken down) and characterized according to this concept. In this case, the figure of the object on the left is triangle. We will now move to the object on the right, and analyze and characterize it in terms of figure, which is triangle. Next, we will divorce ourselves completely from the actual objects and compare the results of the two analyses.

Exhibit #7
Comparing Through Superordinate Concepts

 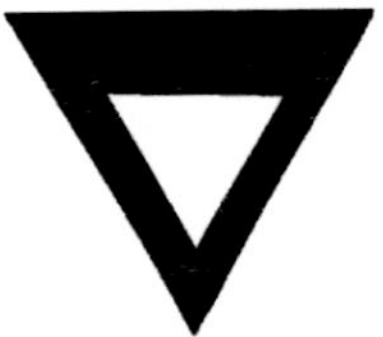

Concept by which to Compare	Object on Left	Object on Right	Similar	Different
Figure				
Shape				
Orientation				
Number				
Quantity (Size)				
Color				
Color Location				
Color Proportion				

Completed Table After Mediation

Concept by Which to Compare	Object on Left	Object on Right	Similarity	Difference
Figure	triangle	triangle	X	
Shape	pointed	pointed	X	
Orientation	upward	downward		X
Number	2	2	X	
Quantity (Size)	large/small	large/small	X	
Color	black and white	black and white	X	
Color Location	black inside white outside	white inside black outside		X
Color Proportion	40% black/60% white	40% white/60%black		X

Triangle is the same as triangle. Therefore, the result of comparing these two objects in terms of the concept of figure is a similarity.

The results from complete mediation of the tasks are shown in the next table. The cognitive tasks in Exhibit #8 are from the instrument Comparisons in the FIE program. These tasks are psychological tools through which we mediate learners to practice comparing through superordinate concepts. For each problem item, learners are to compare the objects by

Exhibit #8

Compare	
1. Common: ____________ Different: ____________	
2. Common: ____________ Different: ____________	
3. Common: ____________ Different: ____________	
4. Common: ____________ Different: ____________	

listing the superordinate concepts they have in common and the superordinate concepts through which they differ. For example, in the first item, there are two triangles. The superordinate concepts that they have in common are figure; shape; color; spatial orientation; number; and vertical location on the page. The superordinate concepts through which they differ are size and horizontal spatial location.

In the third task, a learner must choose one of two options prior to performing the task. The first option is to consider each group of objects as a single whole. One's decision establishes the logical basis, frame of reference, or perspective for defining the whole. With each group as the entire object, the superordinate concepts that they have in common are number, color, and vertical location. They are different based on shape, figure, and horizontal spatial location. The second option, in which each individual item in a group is the whole, the superordinate concepts they have in common are figure, shape, size, color, and vertical spatial location. They differ by number and horizontal spatial location.

These tasks for comparing through superordinate concepts can be highly rigorous and demand abstract conceptual thinking.

Defining the Problem, Conserving Constancy, Forming Relationships, and Hypothetical Thinking

a. Defining the Problem and Conserving Constancy

The set of cognitive tasks shown in Exhibit #9 is from the instrument Orientation in Space I of the FIE program. Learners ask several questions when presented this set of tasks. "Is there a problem here? What are we to do? What is missing?" These questions express the need for the cognitive function **defining the problem.** The mediator responds, "You have on this page a lot of information or data. Let us call this entire page a field of data. This field of data is complex because there are many parts. What cognitive function may we use to begin to address this complexity?" Learners immediately say "Analyzing."

From analyzing the complete field of data, they determine that there are four large frames, with each frame containing four objects—a house, a tree, flowers, and a bench—with an empty rectangle next to each object, and a boy or man. By comparing the four frames, they discover that everything remains the same in each frame except the position of the person.

The mediator points out that the learners have derived the meaning of a new cognitive function **conserving constancy** by analyzing the field of data and comparing the four frames, to determine how they are similar and how they are different. They are directed to focus specifically on what is really staying the same as they move from frames A–D (the four objects and their spatial locations and the empty rectangle next to each object) and then specifically what is different (the spatial orientation of the person). This leads learners to clearly and precisely define the changes that are occurring. It also stimulates within them the emergence of "keeping something the same in one or more ways" as the exclusive meaning of the cognitive function conserving constancy.

Three questions are now posed by the mediator:

Exhibit #9

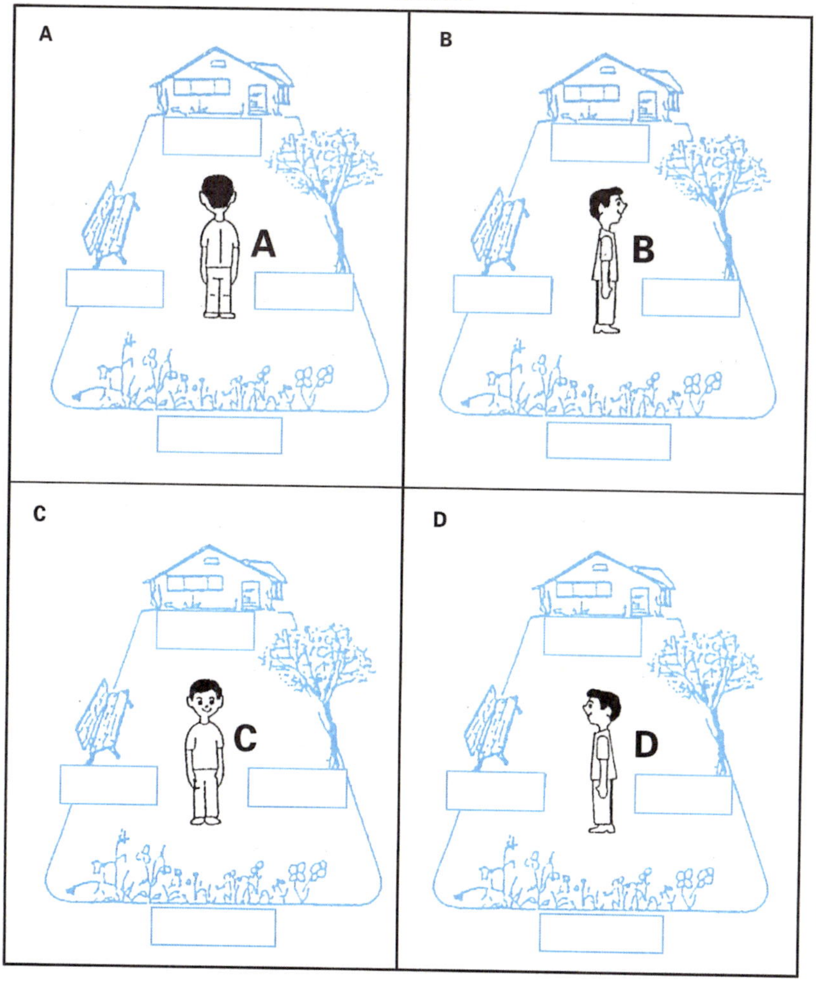

1. What is clearly and precisely known?
2. What is clearly and precisely unknown?
3. How do you logically move from the known to the unknown? What is the logical pathway?

Seeking to answer these questions are three prongs of actions the learners must take in defining the problem.

Forming Relationships and Hypothetical Thinking

Now a source of discrepancy presents itself—the need for a tool and the language to help the learner articulate space and spatial relationships. Again, we turn to Vygotsky's concept of psychological tools, which act as sociocultural mediators to produce a higher level of required mental functioning, abstract relational thinking. Learners are mediated to go beneath the surface and seize each internal reference system as a thinking tool. This mediation commences with creating a more immediate need by giving an instruction like "Point toward your home or some prominent place in the area." It is followed by the question "How do you know it is in this direction?"

Once the need is experienced, the mediator directs attention to the diagram shown in Exhibit #10. Learners are instructed to imagine themselves standing at the center of this representation with the code F as his or her front.

They are then mediated to:

- analyze the structure of the diagram in terms of spatial directions surrounding the center;
- discuss all of the interactions and connections among the parts of these directions;
- build spatial language and concepts to label positions and spatial relationships;
- identify the functions or purposes of this structure; and
- begin perceiving this structure as a relative but stable internal spatial reference system.

With the appropriation of this internal reference system, spatial concepts, and verbal tools, learners construct a logical pathway from the known to the unknown.

Solving the problem requires the learner to superimpose (align) his or her internal reference system (with) on the boy's internal reference system. Next, as in Frame A, the learner engages in hypothetical thinking by saying:

I select the relevant cue that the house is in front of me. By hypothetical thinking, if the house is in front of me, then my internal reference system informs me that the flowers are at my back, providing the logical evidence that my front is always forming an opposite

Exhibit #10

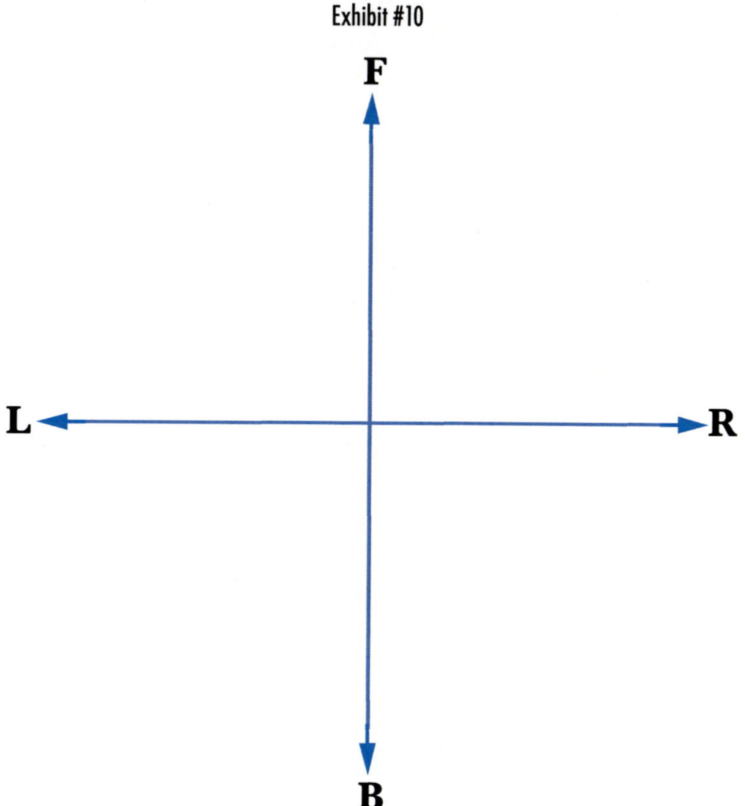

relationship with my back. If the house is at my front, then the tree is at my right, providing the logical evidence that my right is 90° clockwise from my front.

Mediating learners in performing this set of cognitive tasks engages them in beginning to build the meaning of each of the cognitive functions:

- **Forming relationships** as— bringing about an interaction or bonding between things or making connections; and
- **Hypothetical thinking** as—"If ... then thinking" or establishing a hypothesis and searching for logical evidence to support or verify the hypothesis.

How the Use of One Cognitive Function Requires the Use of Three Other Cognitive Functions

How does the cognitive function selecting relevant cues require the use of the three cognitive functions: providing logical evidence, forming relationships, and comparing?

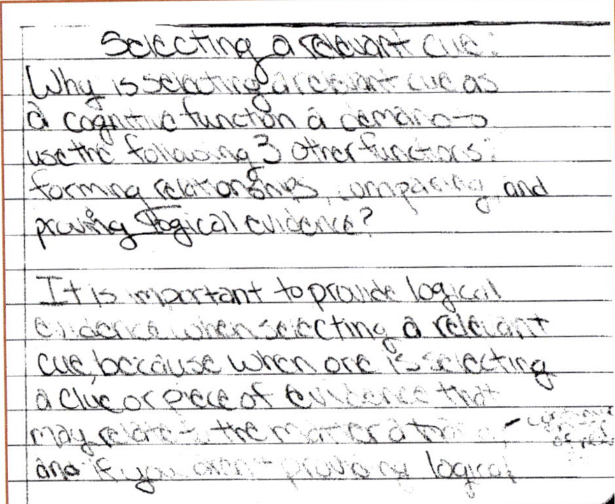

Figure 4.8: A fourth-grade learner describes how selecting relevant cues demands the use of three other cognitive functions

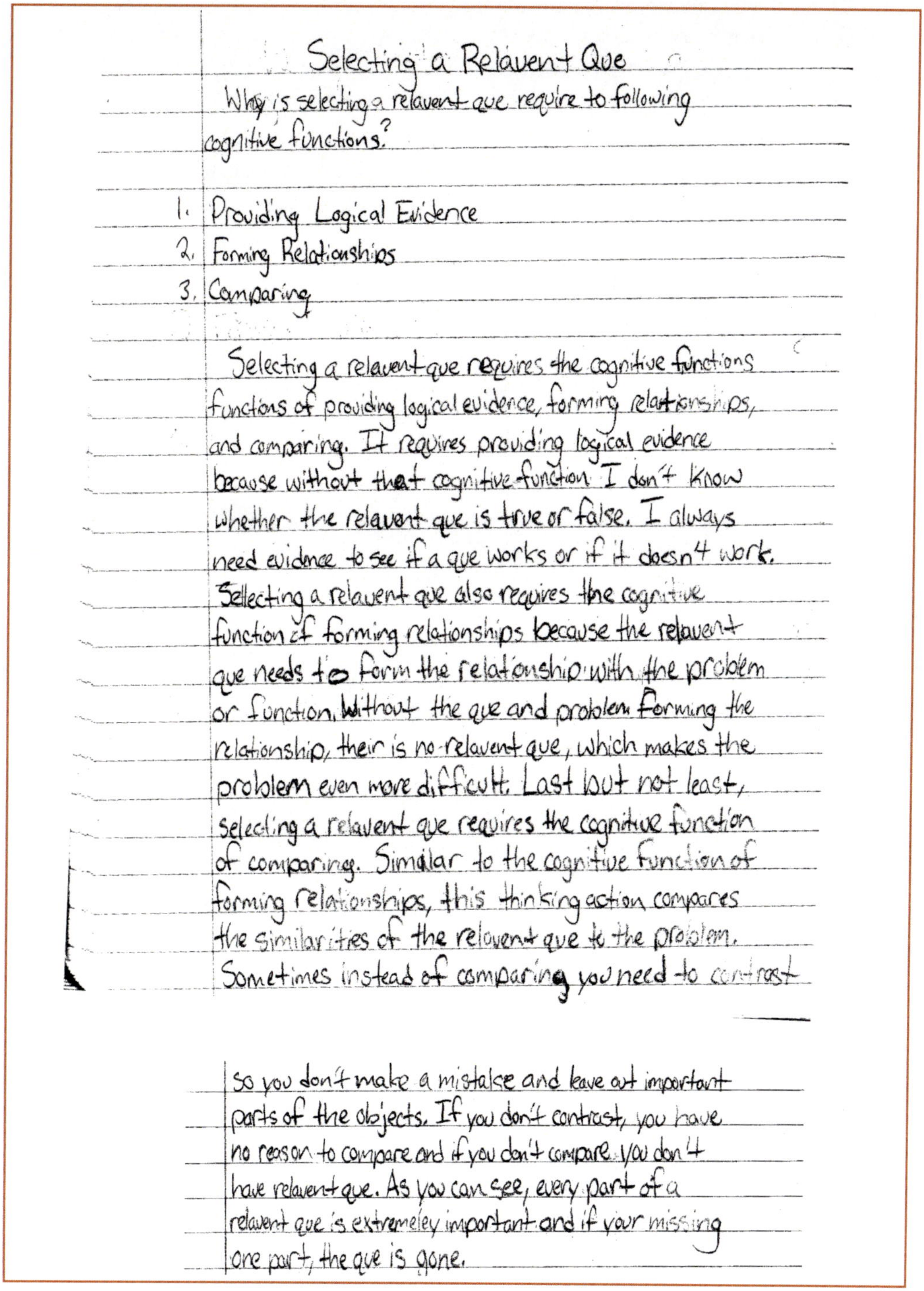

Selecting a Relavent Que

Why is selecting a relavent que require to following cognitive functions?

1. Providing Logical Evidence
2. Forming Relationships
3. Comparing

Selecting a relavent que requires the cognitive functions functions of providing logical evidence, forming relationships, and comparing. It requires providing logical evidence because without that cognitive function I don't know whether the relavent que is true or false. I always need evidence to see if a que works or if it doesn't work. Selecting a relavent que also requires the cognitive function of forming relationships because the relavent que needs to form the relationship with the problem or function. Without the que and problem forming the relationship, their is no relavent que, which makes the problem even more difficult. Last but not least, selecting a relavent que requires the cognitive function of comparing. Similar to the cognitive function of forming relationships, this thinking action compares the similarities of the relavent que to the problem. Sometimes instead of comparing you need to contrast

so you don't make a mistake and leave out important parts of the objects. If you don't contrast, you have no reason to compare and if you don't compare you don't have relavent que. As you can see, every part of a relavent que is extremeley important and if your missing one part, the que is gone.

Figure 4.9: A fifth-grade learner describes how selecting relevant cues demands the use of three other cognitive functions

Applying Cognitive Functions to Form Quantitative Relationships

a. Analyzing-Integrating Quantities in a Linear Fashion

Here we are presenting a new cognitive function, not simply a combination of the two cognitive functions, analyzing and integrating. As discussed under the section "The Anatomy of a Cognitive Function" in this book, and by Kinard and Kozulin (2008, p. 84), a cognitive function has a conceptual component, which provides a steering mechanism of conceptual know-how to the function. When analyzing, the specific mental activity is breaking down or decomposing by moving from whole-to-parts, entire-to-segments, complete-to-portions, full-to-pieces, total-to-each. When integrating, the action is merging or composing by moving from parts-to-whole, segments-to-entire, etc. The concepts of parts-to-whole and whole-to-parts comprise the conceptual component of the cognitive function analyzing-integrating. This conceptual component directs a mental action that is a dynamic spectrum of conceptual opposites.

The set of cognitive tasks in Exhibit #11, from the instrument Analytic Perception of FIE program, provides practice in analyzing-integrating, which is the cognitive conceptual structure of adding-subtracting. Learners are mediated to begin analyzing the entire field of data, reading the instructions at the top of the page, and defining the problem. The following describes one learner's reflection after completing all the tasks.

I analyzed the full model, which was the figure at the top of the page, by breaking it down into two parts, the selected part given from the column on the left and

Exhibit #11

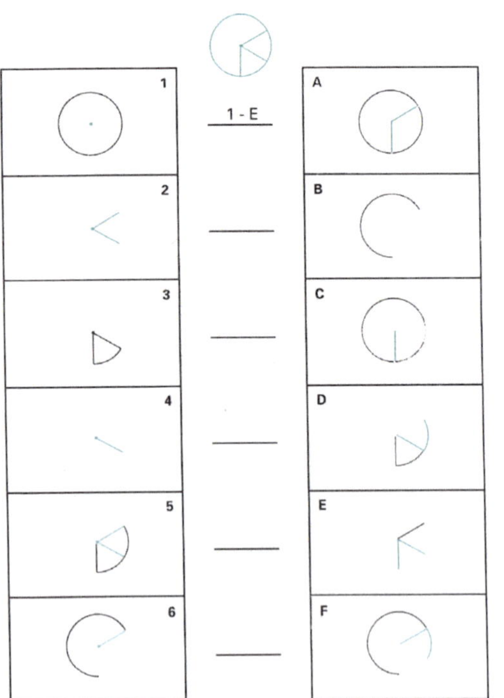

Table 4.1: Building Cognitive Conceptual Structure of Analyzing-Integrating

Task No.	Whole Model	Minus	Drawing in Left Column	Is Equal To	Drawing in Right Column
1	Complete Figure	—	Circle with center point	=	3 radii E
2	Complete Figure	—	Two radii	=	Circle with radius C
3	Complete Figure	—	Arc (2 radii + 1/6 of circumference)	=	Radius + 5/6 of circumference) F
4	Complete Figure	—	Radius	=	Circle + 2 radii A
5	Complete Figure	—	Arc + radius	=	2/3 circumference B
6	Complete Figure	—	2/3 circumference + radius	=	Two radii + 1/3 circumference D

the unknown or missing part that appeared in the column on the right. But as I tried to identify or determine the unknown part, I was steadily integrating what appeared to be the missing part with the given part to provide my logical evidence that this was the correct part. All the time I was moving back and forth comparing the parts I was considering at that moment. In some ways I was selecting differences as relevant cues and doing some hypothetical thinking. I had to always conserve constancy in the model.

The generalized structure of the process is whole figure—the selected part from left column = the missing part from right column. The specific structure for each task is given in Table 4.1.

In Exhibit #12, we are depicting the generalization of analyzing and inte-grating quantities in a linear fashion. Here, we have two sets of schematics or illustrations that capture and express mathematical events.

On the left side we have a diagram, and on the right side we have a set of equa-

Exhibit #12

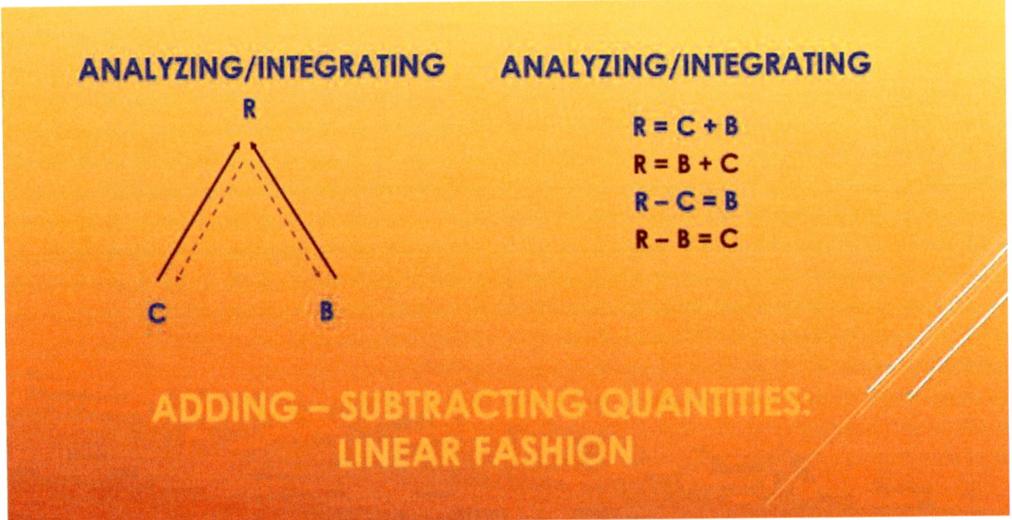

tions. We are analyzing and integrating in a linear fashion, that is without changing any units of measure.

On the left, we are analyzing the quantity R by decomposing it into two parts, the quantity C and the quantity B. Thus, R, B, and C are quantities with the same unit of measure. This mathematical event is captured on the right side by the last two equations. The quantity R is analyzed into two parts, the given quantity C and the unknown quantity B. In the very last equation, we are analyzing the quantity R by breaking it down into two parts, the given quantity B and the unknown or missing quantity C. Again, we are analyzing in a linear fashion because the quantities R, B, and C have the same unit of measure.

Let us now return to focusing on the sche-

matic on the left. As we move upward, we are integrating or merging together the quantities C and B to compose the quantity R. This quantitative composition is depicted in the top two equations on the right side. We are integrating the quantity C with the quantity B to compose the quantity R. We have integrated the quantities C and B to form a quantitative relationship R. In the second equation we are integrating the quantity B with the quantity C to produce the quantity R. These two equations illustrate the commutative property of addition, which establishes that the order of adding or integrating the quantities in a linear fashion does not change the sum or the quantitative composition.

Here we are emphasizing, through both the diagram and the set of equations, that analyzing and integrating are not separate, opposite thinking processes, but they are dynamic cognitive functions that are working together in a synergy by complementing and enhancing each other. In this very same way, subtraction and addition are not merely opposite mathematical operations being performed separately or in isolation of each other. In the RMT paradigm, addition and subtraction dynamically work together as a unified cognitive-conceptual structure that enhances and complements the full process. Procedural fluency and computational proficiency produced this way enhances the learner's ability to dynamically add and subtract simultaneously.

In Exhibit #13 below, we present specific examples of analyzing-integrating in a linear fashion. On the left we have $5cm - 2cm = ?$

A learner is using the cognitive function analyzing.

- He or she is analyzing the 5cm into two parts, the given part 2cm and the other part or the missing part 3cm.

The learner is analyzing in a linear fashion because all quantities have the same unit of measure.

On the right side, we have the quantity 6cm + the quantity 4cm =

- The learner is using the cognitive function integrating.
- He or she is integrating the given parts 6cm and 4cm to form the whole quantity 10cm.

Exhibit #13

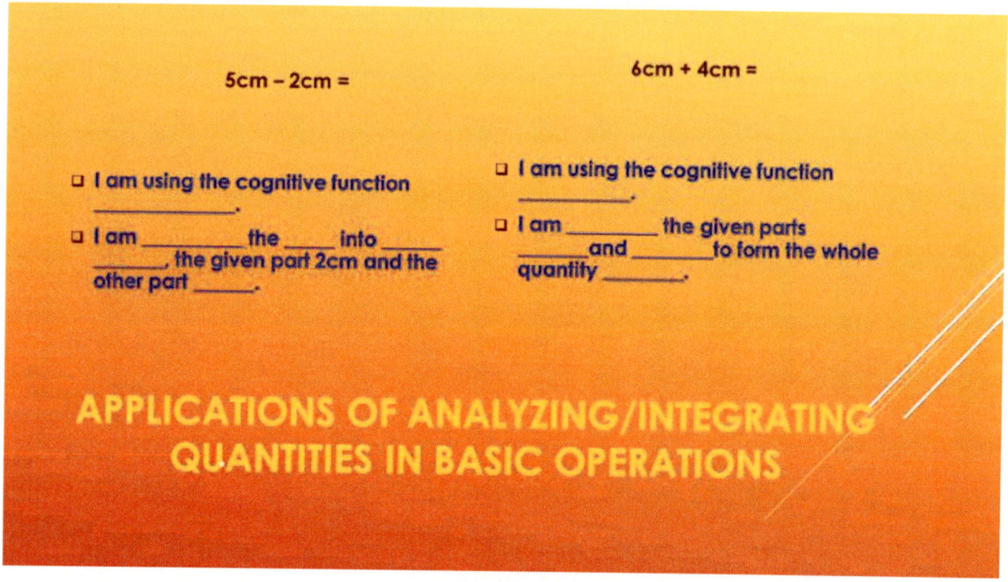

b. Considering More Than One Source of Information Simultaneously
Sets of cognitive tasks that we utilize as general psychological tools to mediate learners in building the cognitive function **considering more than one source of information simultaneously** are presented in Exhibits #14 and #15. These tasks are from the instrument Comparisons of the FIE program.

The written instructions in Exhibit #14 present two sources of information that the learner must consider at the same time to compare accurately. The first source of information is the superordinate category, which captures the similarity of the two missing items, and the second source is "the differences" between the two items. Cognitive functions that the learner must use to support this process are conserving constancy in the conceptual understanding of the similarity, providing logical evidence in producing responses that make sense and forming relationships in making the connections in the similarity of the two items.

In the first task, the learners are mentally guided by the conceptual knowledge of a

Exhibit #14

Fill in two items that correspond to the given similarity and differences.

6

	=	≠
1.	polygon	4 sides / 3 sides
2.	citrus fruit	orange / yellow
3.	number	divisible by 3 / indivisible by 3
4.	musical instrument	wind instrument / stringed instrument
5.	vehicle	marine / aerial
6.	measuring device	for time / for temperature

Comparisons (Adult Version)

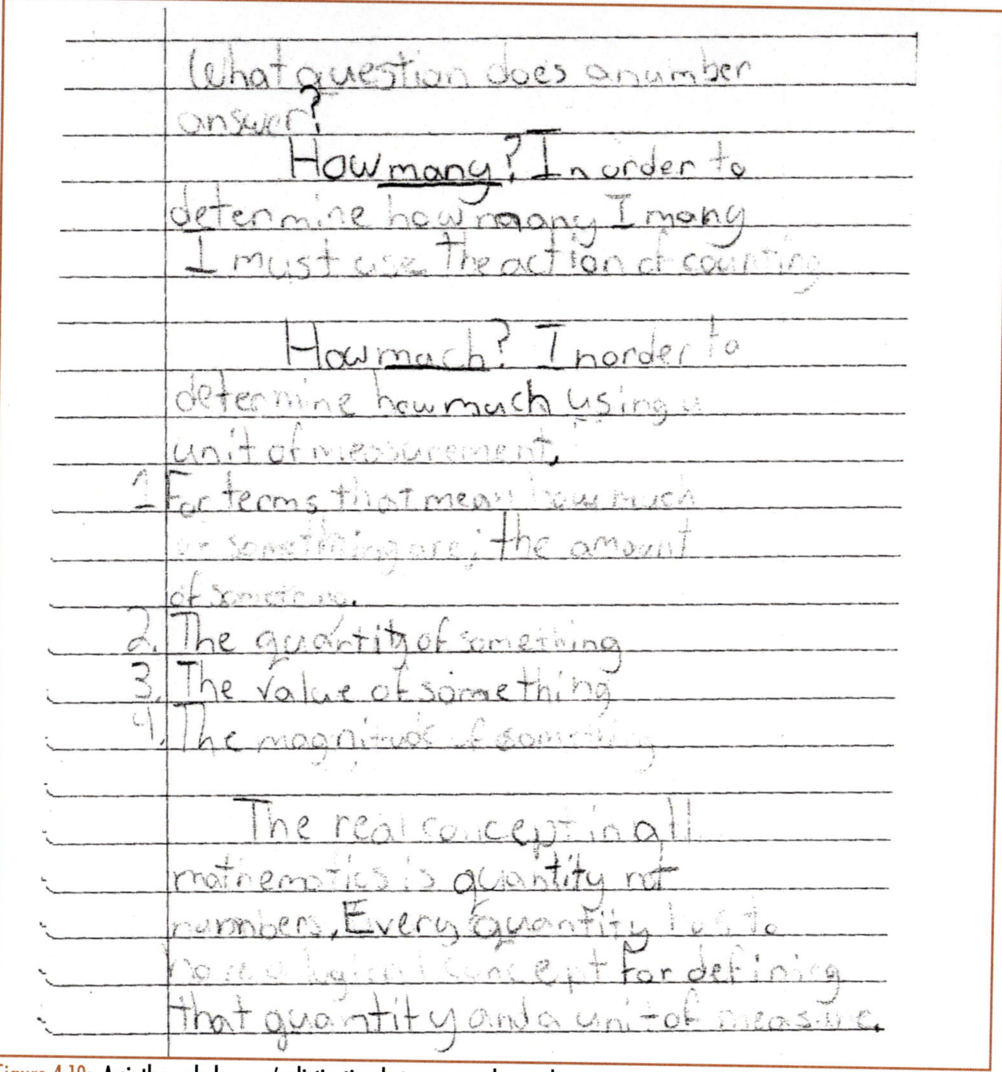

Figure 4.10: A sixth-grade learner's distinction between number and quantity

closed figure with three or more angles, and consequently three or more sides, while at the same time restricting his or her options to those with four sides and three sides. Please notice that each task invokes multiple possibilities in responses, while collectively the set of tasks provides the opportunity for hypothetical thinking and promotes abstract relational thinking.

All these features are essential for rigorous mathematical thinking.

Comparing tasks 3 and 6 provides the opportunity to re-emphasize RMT's distinction between "number" and "quantity." Figure 4.10 gives a sixth-grade learner's analysis and discussion of the two.

Task 6 presents the superordinate concept of "measuring device" or tool for determining quantity. The differences are the corresponding defining concepts for measuring quantity with the different measuring tools. At this point the mediation morphs into a mathematical-scientific discourse. One line of inquiry focuses on the abstract nature of time. Fueled by the mediator's injection: "This reminds me of Albert Einstein's response to the question, 'What time is it?' He stated: 'It's what the clock reads.'"

Discussion, debate, and argument center on various "clocks," from the products of modern technology to an hour glass, the length of a shadow, a person's pulse rate, a rooster's crowing, a swinging pendulum, the ocean's tide, the rings in a tree, the expression of radioactive decay, the thickness of a geological deposit, etc. Emerging from this discourse is a deepening of the understanding that number is a part of the tool system for measuring the quantity of time.

Exhibit #15

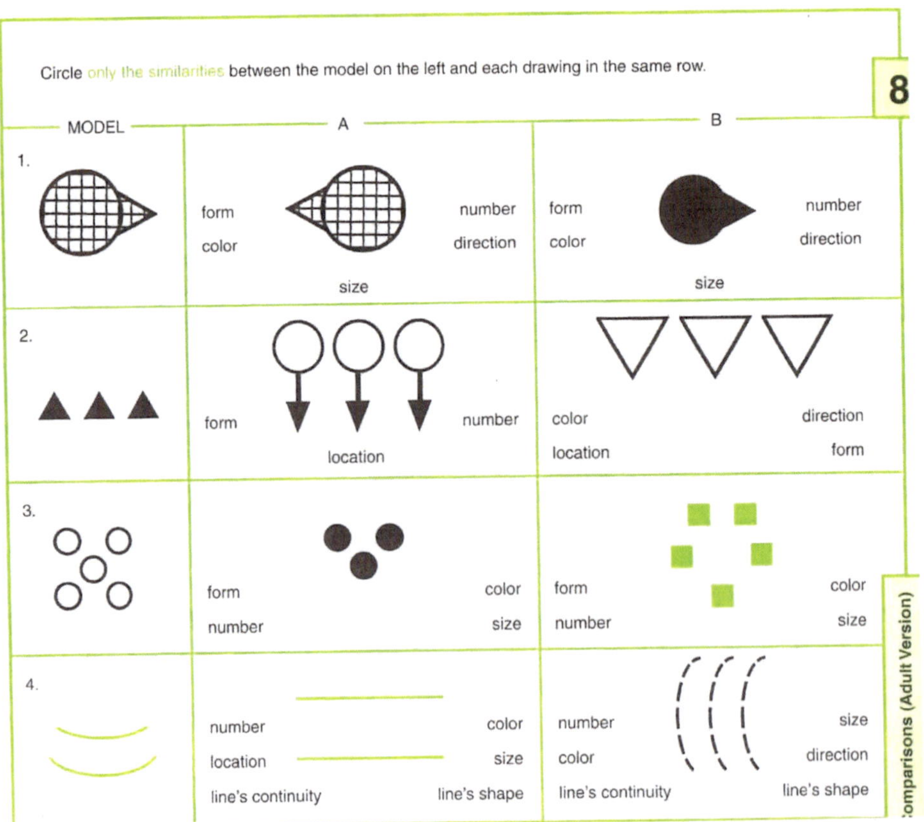

Exhibit #17

12 mi. ÷ 4 s. =

❏ I am using the cognitive function

___________.

❏ I am _________ the quantity _____ into

____

❏ and each part will be _____.

❏ 12 mi. ÷ 4 s = _________

3 ft. X 4 ft. =

❏ 4ft + 4ft + 4ft = _________

❏ 4ft x 3ft = _________.

APPLICATIONS OF ANALYZING/INTEGRATING QUANTITIES IN BASIC OPERATIONS

3 miles per second.

- 12 mi ÷ 4 s = 3 mi/s.

On the right, learners are multiplying 3ft by 4ft. They are forming the relationship that the new unit of measurement will be ft^2. Thus 3 ft x 4 ft translate to integrating 4 sets of 3 ft^2 to form the quantitative relationship, 12 ft^2. Learners are integrating in a nonlinear fashion, with a change in unit of measurement. Compare that with the linear integration of

- 4ft + 4ft + 4ft = 12 ft. This comparison provides mathematical logical evidence that multiplication is not simply repeated addition.
- 4ft x 3ft is integrating 3 sets of 4ft^2 to produce the quantity 12 ft^2.

In Exhibit #18 on the left side is a mathematical event that is presented horizontally: 3 gal 2 qt − 1 gal 3 qt. Learners are analyzing the quantity 3 gal 2 qt into two parts, 1 gal 3 qt and the missing or unknown part.

On the right side, the same mathematical event is presented vertically. There is a discrepancy or disequilibrium in attempting the subtraction of 2 qt − 3 qt. To address this discrepancy, learners need to engage in quantitative conceptual thinking to transform the quantity 3 gal 2 qt without undergoing a change in the units of measure.

When learners analyze the quantity one gallon into 4 equal-sized parts, one of these equal-sized parts is 1 quart. Forming quantitative relationships: They show that 1 gallon is equivalent to 4 quarts and 1 quart is equivalent to ¼ gallon. Thus, they are analyzing 2 gal 6 qt into two parts, 1 gal 3 qt and 1 gal 3 qt. Learners are analyzing in a linear fashion because they are not changing the units of measure (see Exhibit #19).

Exhibit #18

3 gal. 2qt. – 1 gal. 3 qt. =

I am analyzing the quantity 3 gal. 2 qt. into two parts, 1gal. 3qt. and the missing or unknown part.

3 gal. 2 qt.

-1 gal. 3 qt.

When I analyze the quantity 1 gal. into 4 equal-sized parts, one of the equal-sized parts is 1 quart. Forming Relationships:

1 gal. = 4qt.   1 qt. = ¼ gal.

QUANTITATIVE CONCEPTUAL THINKING

Forming Proportional Quantitative Relationships (Ratios, Rates, Proportions, Fractions)

Building the Cognitive Conceptual Structure

The underlying cognitive conceptual structure for forming proportional quantitative relationships is abstract representational-relational thinking. Mediating learners to perform the tasks shown in Exhibit #20 builds this kind of thinking. The tasks are from the instrument Orientation in Space I of the FIE program. The mediation begins with analyzing the entire field of data presented. The results of this analysis are a smaller field of four objects—a house, a tree, flowers, and a bench; a legend or key with four different spatial positions of a man; and a table. At this point, the mediation turns to engaging learners in appropriating a table as a general psychological tool. Typically, in mathematics education and other disciplines, a table is not explicitly distinguished from its content. Its instrumentality as a general tool of thought is often not recognized and used to enhance learning. But the RMT paradigm mediates learners to acquire and operate through the mathematics culture. The structural features of the symbolic artifact of a table lend themselves to the appropriation and internalization of this structure into a dynamic mathematically specific psychological tool to deepen and enhance the quality of mathematical thinking and conceptual learning.

The primary goals of this appropriation are for learners to do the following:

1. recognize that every table has the same basic structure;
2. internalize this structure;

Exhibit #19

3 gal. 2qt. – 1 gal. 3 qt.
= 2 gal. 6 qt. – 1 gal. 3 qt.
= 1 gal. 3 qt.

❑ Analyzing
❑ Forming Relationships
❑ Integrating
❑ Labeling
❑ Visualizing

QUANTITATIVE CONCEPTUAL THINKING

3. mentally seize this structure to carry out specific functions; and,
4. practice using different tables as thinking and learning tools.

In RMT, structure is the interrelatedness of the parts or stages of an object or operation. The structure of a table begins with its columns, which run vertically, and its rows, which run horizontally. Each column has a heading, which serves as a superordinate category that organizes all items under it into a family or a set. Moving from left to right in a single row from column to column brings about the forming a set of relationships.

The first column of the table in Exhibit #20 has the heading "Object," and each cell under it contains a verbal label for one of the objects pictured in the field. The second column has the heading "Side of person" and each cell under it is filled in with "Front," "Left," "Back," "Right," etc.

Applying the cognitive function defining the problem, learners determine that the known comprises the given contents of the field of four objects, the position legend, and the first two columns of the table. The unknown is the missing content of the third column, headed "Position," for each row.

Let us consider the object "Tree" in the second row. A learner may take the following actions:

1) focus on the verbal label "Tree": 2) visualize a tree in his/her mind; 3) systematically search the field of objects while comparing to identify the target object; 4) select "Left" in the next column as a relevant cue; 5) mentally represent himself or herself in the center of the

Exhibit #20

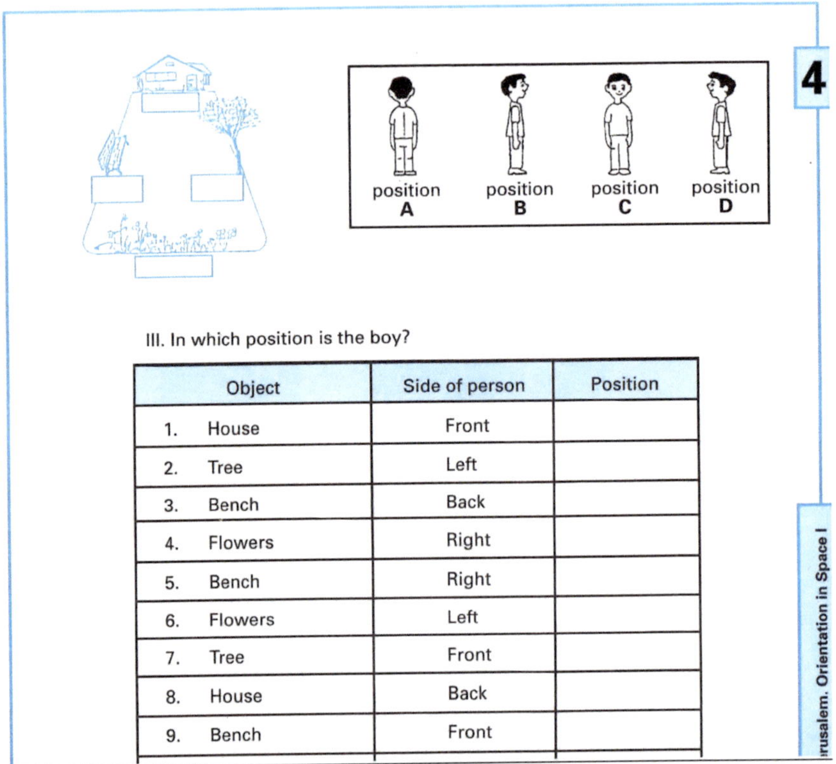

Object	Side of person	Position
1. House	Front	
2. Tree	Left	
3. Bench	Back	
4. Flowers	Right	
5. Bench	Right	
6. Flowers	Left	
7. Tree	Front	
8. House	Back	
9. Bench	Front	

field of objects; 6) use his or her internal reference system as a tool to spatially align his or her orientation such that the Tree is forming a relationship on his or her left side.

The learner must now engage in hypothetical thinking to validate his or her outcome. "If the Tree is on my left, by using my internal reference system, a 90^0 clockwise rotation provides the logical evidence that the Flowers are in front of me, the House is at my back and the Bench is on my right."

Thus, the structure of this table can be utilized to carry out the functions of organizing and forming logical relationships within a highly complex field of data. In this regard, it is important to notice that the high level of abstract representational-relational thinking is promoted by the interfacing and simultaneous use of two general psychological tools, one's internal reference system and the table.

Constructing Quantity of Linear Space

Learners are mediated to use the cognitive function **activating prior knowledge** to connect with and review the concept of quantity. How much of something is the quantity of something. It is the amount, value, or magnitude of an object that is determined from the action of measuring. Next, they are mediated to build the understanding that a point is a location or a position in space. A point is abstract because it cannot be interacted with through any one of the five senses. One can only perceive or interact with a point through use of cognitive functions. We represent a point on paper or on the board with a dot.

The most direct path between two different points in space is a straight or linear path. The quantity of space comprising this most direct path between two points is the quantity of linear space. A line segment consists of two endpoints and contains a certain quantity or amount or value or magnitude of linear space. A line segment is also abstract. We attempt to represent a line segment with a straight mark on a piece of paper or a white board.

Composing a System of Proportional Quantitative Relationships

The mediator gave learners the following instructions:

(a) Construct a horizontal representation of a quantity of linear space (not very long) and encode it E.

(b) Construct a horizontal representation of a second quantity of linear space that is twice the amount of E and encode it K.

(c) Construct a horizontal representation of a third quantity of linear space that is three times the amount of E and encode it G.

(d) Form all possible relationships between E, B, and K with the equal sign by strictly using your cognitive functions and without making substitutions. Express each quantitative relationship in words using complete mathematical language. Provide your mathematical logical evidence to justify your results.

Figures 4.11 and 4.12 show the results of a sixth-grade learner. Notice the learner's explicit use of his cognitive functions to systematically produce his mathematical outcomes.

Thus, the unique, specific meaning of the cognitive function forming proportional quantitative relationships is creating or constructing a system or a network of quantitative correspondence, where two or more quantities are interacting with each other in definite proportions or ratios or size relations, such that any change in the amount of one quantity will spontaneously bring about changes in the other quantities so that their original ratios will always conserve constancy.

Concept of Color Proportions

One of the objects learners compared

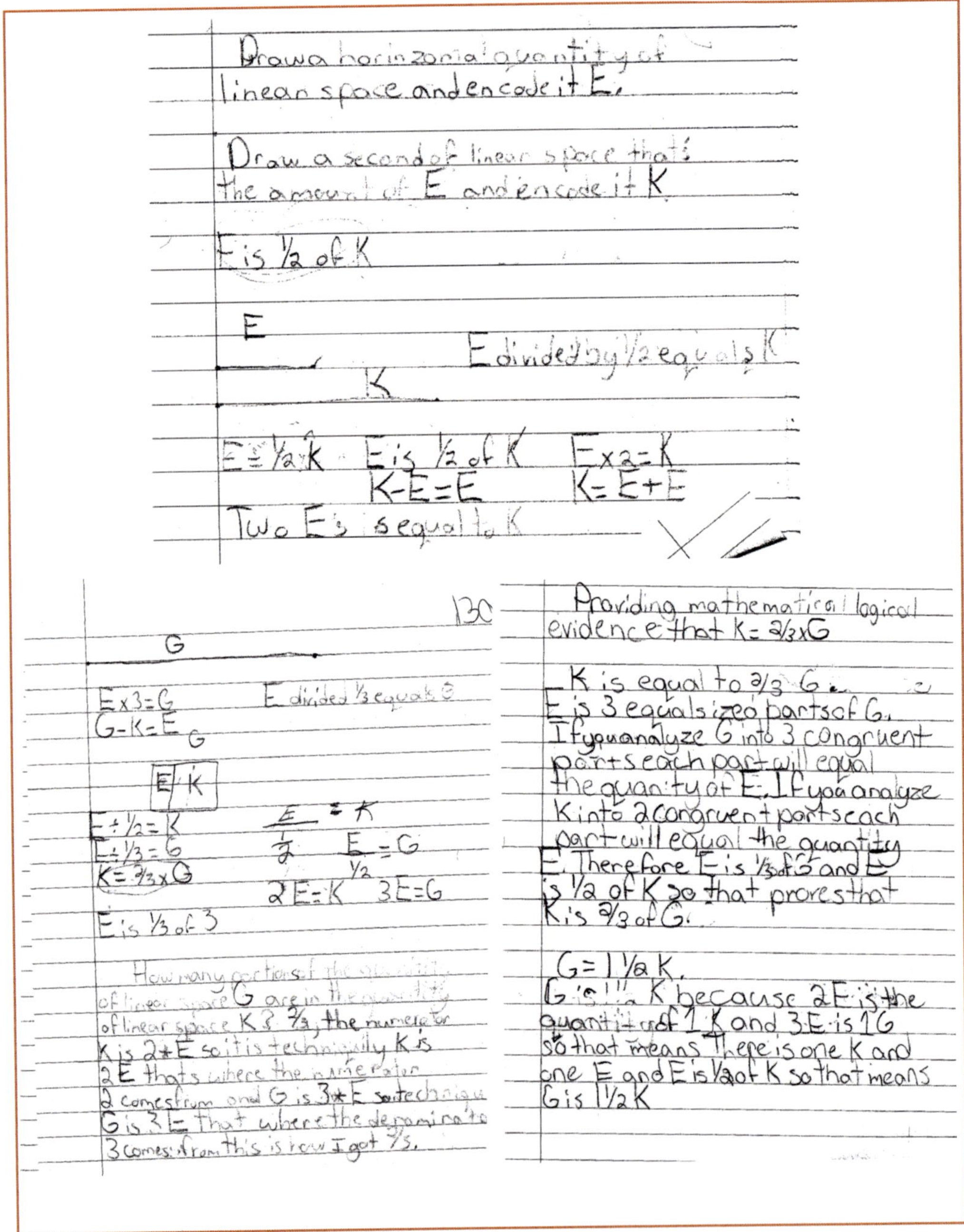

Figure 4.11: A sixth-grade learner's composition of a system of proportional quantitative relationships

the cognitive demand
in providing mathematical
logical evidence

I used the following cognitive
functions to provide mathemat
ical logical evidece. logical
evidence selecting a relevant
cue analysing hypothetical
thinking comparing forming
a relationship, and conserve
constancy. I provided logical
evidence because I showed
I selected a relevant cue
by using the lines as my
main proof. I analyzed by
by breaking down the problems
The hypothetical thinking
because I got a hypothis
and prove it. I comparing
by telling how much some
thing is like something else.
I formed a relationship

Figure 4.12: A sixth-grade learner derives the meaning of forming proportional quantitative relationships

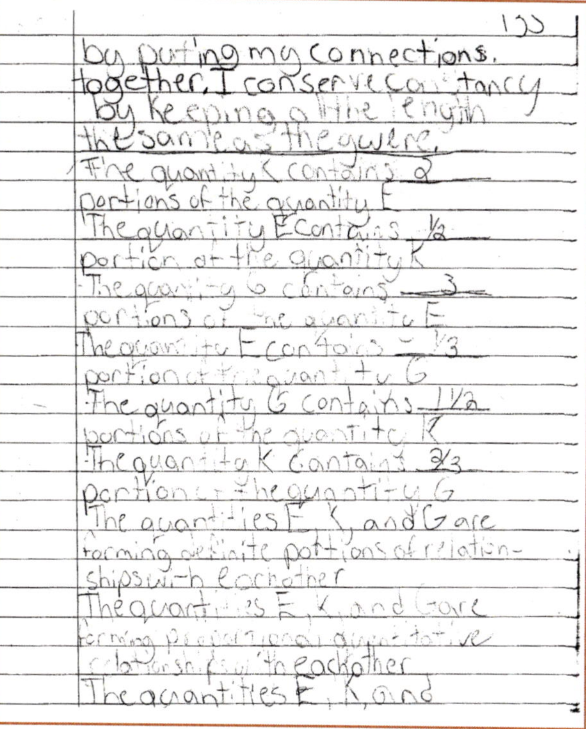

136

G are in a system of definite proportions

This system is like gears the
the smallest but the gear given
power is E. The inner medium gear is
K. The biggest out the one that
moves the conveyor belt is G. Without
K the conveyor belt won't move.

These gears are functioning
together.

E, K, and G are function together
because they have a relationship
E can go into K 2 and K can go into
E ½. E can go into G 3 times and
G can go into E ⅓ time, K can go into

Figure 4.12: A sixth-grade learner derives the meaning of forming proportional quantitative relationships (continued)

IF this is the system
of Proportional Relationships
$\frac{E}{K}$

$\frac{G}{}$

$K = 2E \qquad E = \frac{1}{2}K$
$G = 3E \qquad E = \frac{1}{3}G$

$K = \frac{2}{3}G \qquad G = 1\frac{1}{2}K$

For every K there are two
E's

For every quantity K there
are 2 quantity E's
K to E   Ratio
K:E   K/E
½

138

Conserving constancy
I used the cognitive function-
Forming quatative relationships
quantitative relationships
because I created E, K, and
G

above was a black triangle inside of a white triangle. They estimated the quantity or amount or value or magnitude of the black to be 40% and the amount of the white to be 60% of the total colors. Forming a relationship (ratio) between the quantities of the two colors in the object:

40% black to 60% white

40 black: 60 white
40 black/60 white
For every 60 portions of white there are 40 portions of black.

This expresses the mathemtical concept of a ratio. A ratio comes from forming a relationship between two quantities. Suppose we reduce the size of the object by considering ½ of its original size, while at the same time conserving constancy in the color ratio.

This ratio can now be expressed as 20 black to 30 white or 2 black/3 white. The ratios of black to white are forming a system of proportional quantitative relationships (see chart below).

Building Flexibility in Abstract Representational-Relational Thinking

We deliberately choose tasks that progressively increase in complexity and abstraction as we concentrate on the process of building cognitive functioning in learners. The tasks in Exhibit #21 are designed to build flexibility in learners' abstract representational-relational thinking. These tasks appear in instrument Orientation in Space I of FIE. This highly complex set of tasks exists in five modalities of presentation —graphical, tabular, numerical, verbal, and symbolic. Thus, application of the cognitive functions analyzing and defining the problem, are most essential from the outset.

Analyzing this complex field of data yields the following four components:

1. a 3-by-3 grid with 9 squares that are numbered consecutively from left to right, beginning with the number 1 in the top left and ending with the digit 9 in the bottom right;
2. a written verbal instruction;
3. a table with 4 columns—the first column has the heading "No. of square," the second column has the heading "Location of square wihin grid," the third column has

Multiple of Original Size of Object	1	2	3	5	12	20
Portions of Black to Portions of White	2/3	4/6	6/9	10/15	24/36	40/60

Exhibit #21

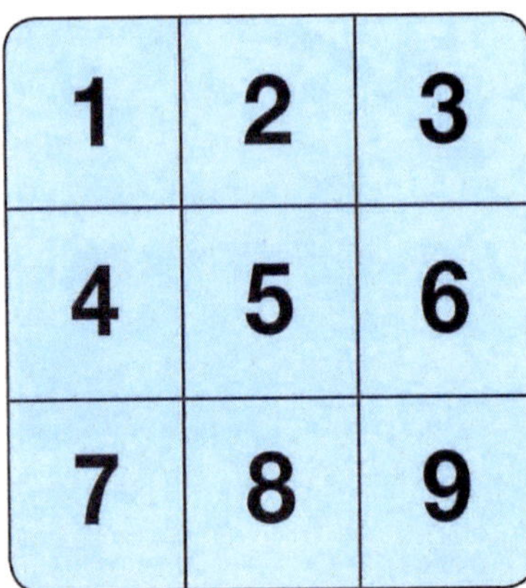

1. In the second column, describe the location of each square within the grid.

No. of square	Location of square within grid	Location in relation to person A	Location in relation to person B
1	Upper left	Front left	
2	Upper middle	Front	
3			
4	Middle left		
5		________	________
6			
7			
8			
9			

Person A is standing in the center of the pattern (in square no. 5), facing square no. 2. In the third column of the table, describe the location of each square in relation to this person.

Person B is standing in the center, facing square no. 9. In the right column of the table, describe the location of each square in relation to this person.

the heading "Location in relation to person A," and the fourth column has the heading "Location in relation to person B"; and,

4. two written verbal instructions.

In the first column, the entry for each row is filled with a digit, beginning with 1–9. Three cells in the second column are completed and two cells in the third column are completed. All of the above comprises the Known in this problem, and the empty cells in the remainder of the table are the Unknown. The third step in applying the cognitive function, defining the problem, requires learners to construct a logical pathway that connects the known to the unknown. The structure and content of the components of the known provide the need to establish frames of reference through which learners can perceive, understand, and manipulate the data. For example, the grid provides a structure that organizes digits into quantitative relationships. Moving horizontally from one square to the next adjacent square, from left to right, C + 1 = number in the square, where C is the quantity in the starting square. Learners are forming a relationship between the quantities C and 1 through integrating them in a linear fashion. Moving vertically, downward, from one square to the next adjacent square, C + 3 =   . Learners are forming a relationship between the quantities C and 3 through integrating in a linear fashion.

The first vebal written instruction requires learners to describe the location of each square within the grid. The three entries in column for squares 1, 2, and 4 respectively, provide the cues "Upper left," "Upper middle," and "Middle left." Selecting these as relevant cues, learners re-analyze the grid into three horizontal spatial regions and label them respectively with the superordinate spatial concepts "upper," "middle," and "lower." They also form three vertical spatial regions, "left," "middle," and "right." After reading and developing an understanding of the first instruction at the bottom of the page, a learner uses this frame of reference, and visually transposes and represents himself or herself as person A, standing on square no. 5 and facing square no. 2. Using his or her internal reference system, the learner engages in hypothetical thinking to form spatial relationships with all the other numbered squares. A learner could say, "If square no. 2 is in front of me, at my back, forming an opposite relationship, is square no. 8. If square 2 is in front of me, then my internal reference system provides the logical evidence that $90°$ counterclockwise on my left is square no. 4, and the opposite relationship to this is my right where square no. 6 is located. Since square no. 1 is between my front and my left, it is at my front left, etc." After reading and developing an understanding of the second instruction at the bottom of the page, a learner participates in representational thinking to visually transport himself or herself to the center of the grid and stands on square no. 5 facing square no. 9. Using his or her internal reference system, the learner engages in hypothetical thinking by mentally saying: "If square no. 9 is in front of me, at my back, forming an opposite relationship, is square no. 1. If square 9 is in front of me, then my internal reference system provides the logical evidence that 90^0 counterclockwise is my left, at square no. 3, and the opposite relationship to this is my right where square no. 7 is located. Since square no. 6 is between my front and my left, it is at my front left, and since squre no. 4 is between my back and my right, it is at my back right, etc.

Performing these tasks, which are embedded in the context of vigorous change, requires the learner to interface the use of the grid, his or her internal reference system, the table, and verbal concepts to analyze, represent, and logically articulate space and spatial relationships. Each learner must dynamically use other cognitive functions to support this development. For example, the learner must spontaneously compare, consider multitudes of sources of information simultaneosly, conserve constancy, and quantify space and spatial relationships.

Constructing Quantity of Rotation of a Line Segment about a Fixed Endpoint (the Measure of an Angle)

The size of an angle is the amount or quantity or value or magnitude of rotation of a line segment from an initial position to a final position (with one endpoint fixed and the other endpoint free to move, while always conserving constancy in the line segment's quantity of linear space). We begin with two locations of the line segment, its initial position, and its final position after a counterclockwise rotation. The initial position of the line segment establishes a frame of reference for measuring the quantity of rotation of the line segment. This line segment is considered to move from the fixed endpoint and extend horizontally to the right. When the quantity of a complete or full rotation is analyzed into 360 equal-sized parts, one of those equal-sized parts is rotation of 1^0 (1 degree). Thus, a full or complete angle contains 360^0 (see Exhibit #22 below). One-half of a full rotation contains 180^0, and the angle is labeled a straight angle. A quantity of rotation greater than 180^0 is labeled a reflex angle. One-fourth of a full rotation is a quantity of 90^0, and the angle with this measure is labeled a right angle. An angle containing a measure of more than 90^0 and less than 180^0 is labeled an obtuse angle. An angle containing less than 90^0 and more than 0^0 is labeled an acute angle.

Geometric Concept of Congruence

The conceptual component of a cognitive function plays a significant role in constructing mathematical conceptual understanding. As discussed earlier, a well-developed cognitive function is a mediated internalized structure of conceptual meaning that equips the learner with the ability and intrinsic motivation to execute a conceptual action. Here we will examine the role of the combined conceptual components of the cognitive functions conserving constancy and forming relationships in constructing conceptual understanding of the mathematics idea of proportionality.

Conserving constancy has the conceptual meaning of "maintain sameness," "invariance," "uniformity," "lack of change," "continued occurrence," "regularity," etc. For forming relationships, the conceptual component of the cognitive function is "bonding," "interaction," "linkage," "connection," etc. Let us examine how these two cognitive functions forge the conceptual understanding of forming proportional quantitative relationships.

We begin with the two quantities, a rope's length and its weight. For example, suppose we have 20m of rope that weighs 1kg. This is 20m of linear space and 1kg of weight. The first representation in Ex-

hibit #23 is expressing an interaction, a linkage, a connection, or a bonding between these two quantities, such that we no longer have 20 m and 1 kg, but we have 20m of linear space for every 1kg of weight, or 20m per kg (20m/kg). Thus, we are forming a quantitative relationship. This bonding expresses the maintenance of sameness, regularity, and uniformity in this quantitative relationship. If one of these quantities changes, then the other quantity must change by a corresponding amount to conserve constancy in this relationship. Thus, if we double the 20m to 40m, we must also double the 1 kg to 2 kg, or if we take 1/5 of 20m, we must take 1/5 of 1 kg such that 4m/0.2kg = 40m/2kg = 20m/1kg, etc. This system of quantitative proportionality captures the concept of a binding uniformity in the quantitative nature of the rope. Thus, we are forming a proportionality when we are always conserving constancy in the quantitative relationship between two objects, the length and the weight, even as the objects themselves change their amounts or values or magnitudes.

The big mathematical conceptual struc-

Exhibit #22

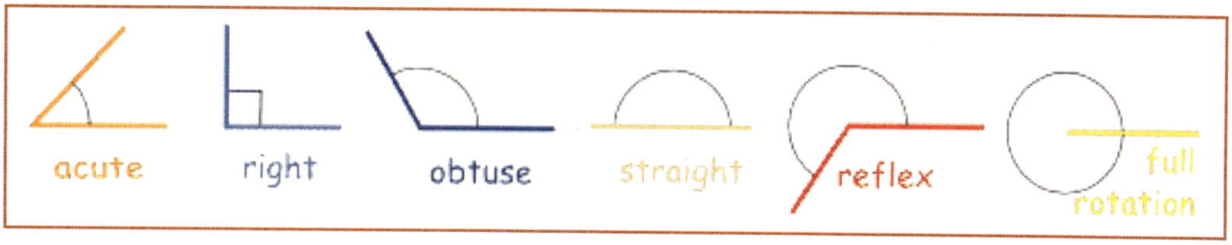

Exhibit #23

ture, **forming proportional quantitative relationships** (which by the way is also a cognitive function), comprises a system of related mathematics concepts. These concepts are ratios, rates, proportions, congruence, similarity, fractions, linear equations, and slopes. This is evidence that these mathematical ideas exist within the same systemic conceptual fabric, which in turn structures the means through which leaners can create their mathematical insights. We will now explore how the dynamic use of cognitive functions and their conceptual components contribute to constructing conceptual understanding of some of these concepts.

First, we turn to a set of cognitive tasks in the FIE instrument Organization of Dots shown in Exhibit #24.

Strategically selected pages of this instrument serve as cognitive tasks through which learners are mediated to build RMT's operational definition of cognitive functions. One cognitive function that is pertinent to each set of tasks is defining the problem. In every situation or set of conditions, learners must go beneath the surface to clearly and precisely identify and describe the source and nature of discrepancy in a field of data and determine its cause, and then design one or more logical pathways to overcome the discrepancy and establish equilibrium. This ongoing approach cultivates within learners the orientation and need to analyze and transform any situation or set of circumstances into a problem that must be defined with the potential for solution pathways that must be invented and evaluated.

In a similar fashion, the RMT approach utilizes a set of general psychological tools to build cognitive functions and concomitantly transpose these tasks into mathematical events. Clear evidence of this phenomenon is presented in Figure 4.13.

The learner explicitly and vividly constructs the geometric concept of congruency through the conceptual meaning and action of the cognitive function conserving constancy:

Exhibit #24

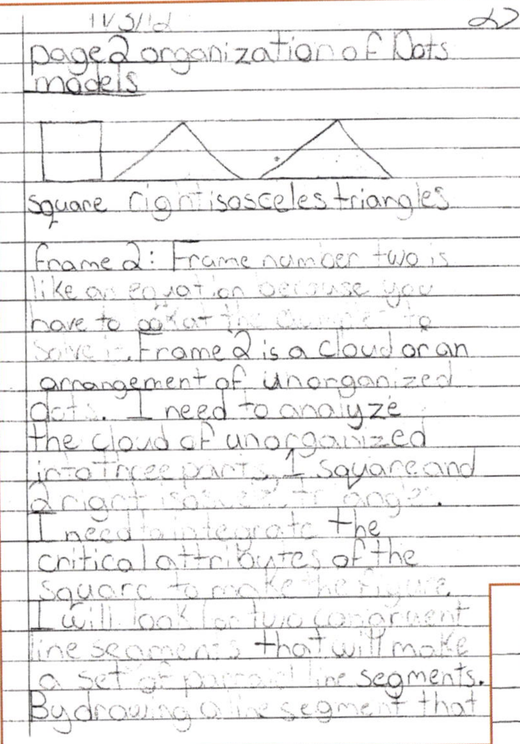

Figure 4.13: A sixth-grade learner's reflections on how he performed the cognitive tasks in Exhibit #24

the learner expresses this conceptual understanding: "I need to integrate the critical attributes of the square to make the figure."

Further, the learner expresses conserving constancy as a key conceptual attribute of parallel line segments. He writes: "Equidistant means that the two line segments have the same spatial separation."

Throughout the cognitive process of projecting virtual relationships

I need to organize the 6 unorganized dots into two parts, two right isosceles triangles that conserve constancy in size and shape with the model, whereby they will be congruent with each other.

In geometry, two objects or figures are congruent if they have the exact same shape and size. Through RMT, learners discover that the shape or form or structure of a geometrical figure is produced from its properties or critical attributes. Notice that

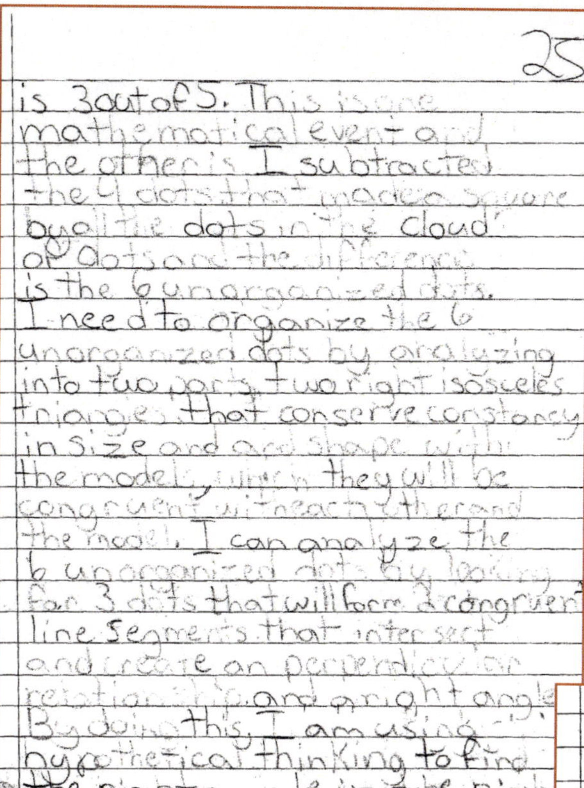

Figure 4.13: **A sixth-grade learner's reflections on how he performed the cognitive tasks in Exhibit #24 (continued)**

within a cloud of unorganized dots, he defines the cloud of dots as his whole or his "frame of reference," which he analyzes into parts and forms relationships among the parts. In this regard, he writes: "So now 10/10 angles were organized or I would say 1 whole of the frame was organized." He readily and fluently expresses the mathematical events involving parts-to-whole relationships, proportional quantitative relationships, ratios, and fractions as he reflects on his performance of the cognitive tasks.

Applying Cognitive Functions to Construct Conceptual Understanding of Fractions

What is a fraction? When this question is posed to an RMT learner, he or she activates the cognitive function defining the problem, which requires him or her to become oriented to the situation, information, and data embodied in the question. In this case, a learner selects "fraction" as a relevant cue and turns to establishing a frame of reference or perspective from which to consider the inquiry.

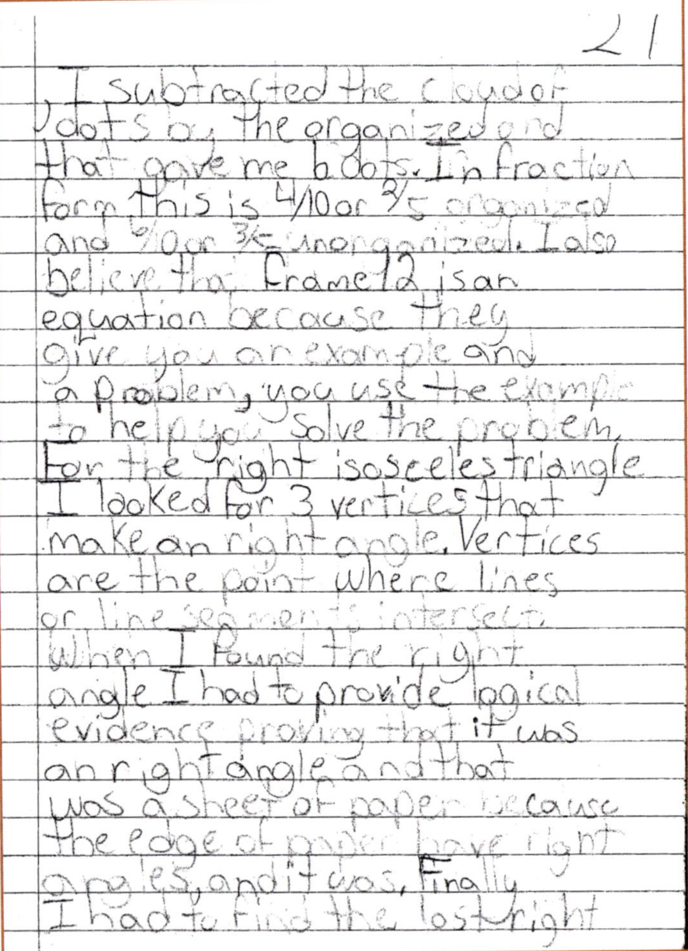

Figure 4.13: A sixth-grade learner's reflections on how he performed the cognitive tasks in Exhibit #24 (continued)

construct of the "full or entire amount of something." For example, it could be the population of a city, a person's paycheck, the duration of a chess game, the size of the entire class, etc. All he does in answering the question is through this perspective. First, he analyzes the "whole quantity of something" into equal-sized parts. Next, he labels the parts, forms relationships among the parts, and then integrates the parts to recompose the "whole quantity."

A fourth-grade learner's response to the opening question is given in Figure 4.14. Notice that the logical frame of reference the learner establishes is "a whole or complete quantity of something." It is very important to realize that the learner is not speaking of a whole number, but the generalized theoretical

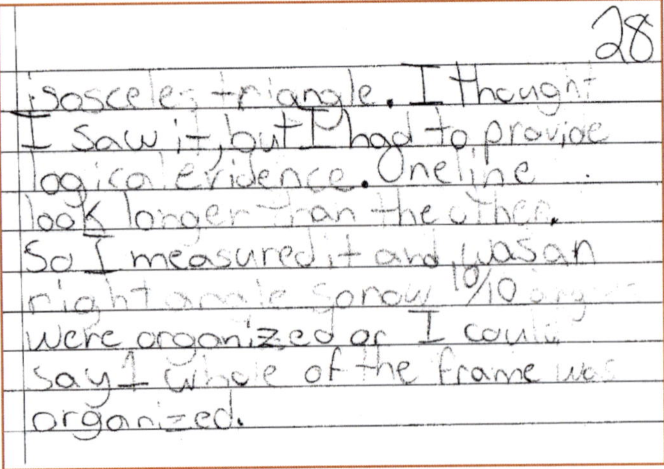

Figure 4.15 presents a fifth-grade learner's responses during an assessment of his conceptual understanding of fractions. Notice that in item #1, when asked to describe the meaning of 4/9 and 3/5, he operates from the framework of the complete or full quantity of "one."

In Figure 4.16, we present a fourth-grade learner's performance on this same assessment. He provides his mathematical logical evidence for how he size-ordered the fractions using abstract representational-relational thinking shown in the second panel of the presentation.

In the third panel of his display, he uses abstract representational-relational thinking to perform the operations involving fractions. He demonstrates the first mathematical event 1/3 + ½. He begins by drawing a circle to establish his frame of reference that represents the complete or entire amount of something. Next, he analyzes this representation into 3 equal-sized parts and shades one of the parts to express the fraction 1/3. He now draws a circle that is, for the most part, identical to the first circle. He uses the same frame of reference —the same "whole quantity of something," re-analyzes it into 2 equal-sized parts and shades one of these parts to represent the fraction ½. He brings the two fractions into the same logical framework through which he can form meaningful quantitative relationships.

To further establish a pathway for the bonding of the two fractions, he re-analyzes the whole quantity into 6 equal-sized parts. He shows that 1/3 is equivalent to 2/6 and that ½ is equivalent to 3/6. Thus, 1/3 + ½ = 5/6 through a linear integration of these two quantities.

Problem 4b is ½ ÷ 1/3. Now the learner establishes a new orientation by considering that he is analyzing ½ into thirds. Therefore, he poses the

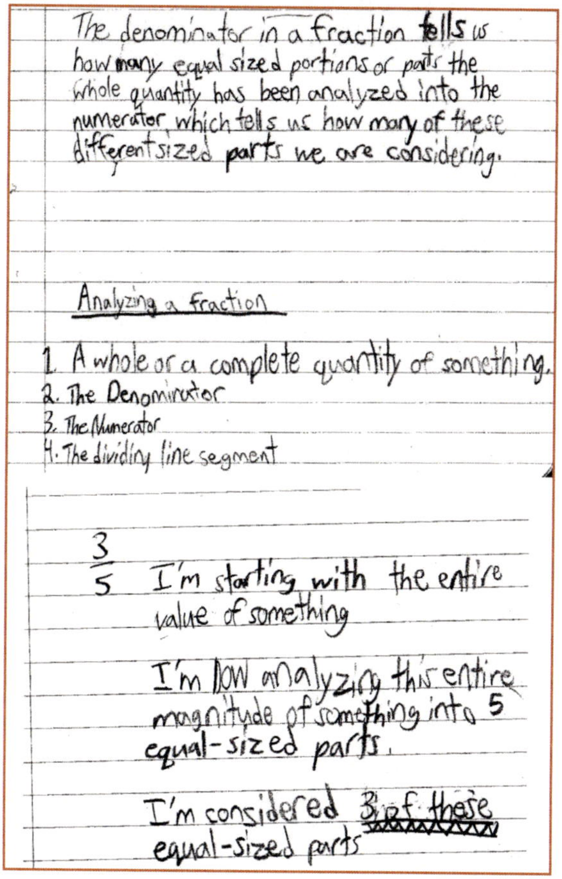
The denominator in a fraction tells us how many equal sized portions or parts the whole quantity has been analyzed into the numerator which tells us how many of these different sized parts we are considering.

Analyzing a Fraction

1. A whole or a complete quantity of something.
2. The Denominator
3. The Numerator
4. The dividing line segment

3/5 I'm starting with the entire value of something

I'm now analyzing this entire magnitude of something into 5 equal-sized parts.

I'm considered 3 of these equal-sized parts

Figure 4.14: Fourth-grade learner's application of cognitive functions in understanding the concept of a fraction

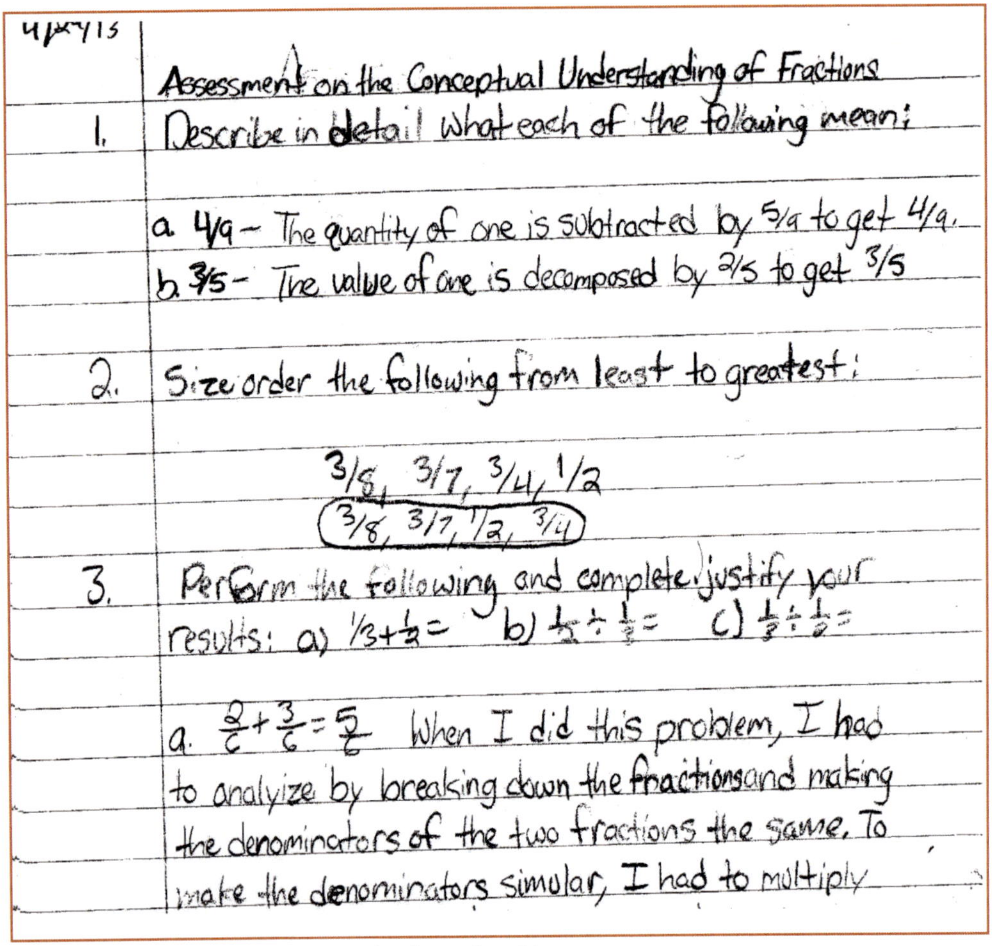

> 4/24/15
>
> Assessment on the Conceptual Understanding of Fractions
>
> 1. Describe in detail what each of the following mean:
>
>   a. 4/9 – The quantity of one is subtracted by 5/9 to get 4/9.
>   b. 3/5 – The value of one is decomposed by 2/5 to get 3/5
>
> 2. Size order the following from least to greatest:
>
>   3/8, 3/7, 3/4, 1/2
>   3/8, 3/7, 1/2, 3/4
>
> 3. Perform the following and complete. Justify your results: a) 1/3 + 1/2 =   b) 1/2 ÷ 1/3 =   c) 1/3 ÷ 1/2 =
>
>   a. 2/6 + 3/6 = 5/6   When I did this problem, I had
>   to analyize by breaking down the fractions and making
>   the denominators of the two fractions the same. To
>   make the denominators simular, I had to multiply

Figure 4.15: A fifth-grade learner's conceptual understanding of fractions

question: "How many 1/3's are in 1/2?" His quantitative thinking inspires him to construct another question to solidify his frame of mathematical reasoning: "Will the result be less than or greater than the original whole quantity?" He reasons since $1/2 > 1/3$, the result > 1.

This gives him the orientation to engage in inferential-hypothetical thinking: "If I have 2/2, how many one-thirds will I have? Of course, there are 3 one-thirds in 2/2. But I really have ½ instead of 2/2. Therefore, my answer will be one-half of 3 or 3/2. Thus, $½ ÷ 1/3 = 3/2$. There are one and one-half one-thirds in ½. Problem 4c is $1/3 ÷ ½$. The learner follows the same scenario of mathematical reasoning and concludes that the result will be less than 1. In his inferential-hypothetical thinking: "If I have three 1/3's, how many ½'s will I have? For sure I

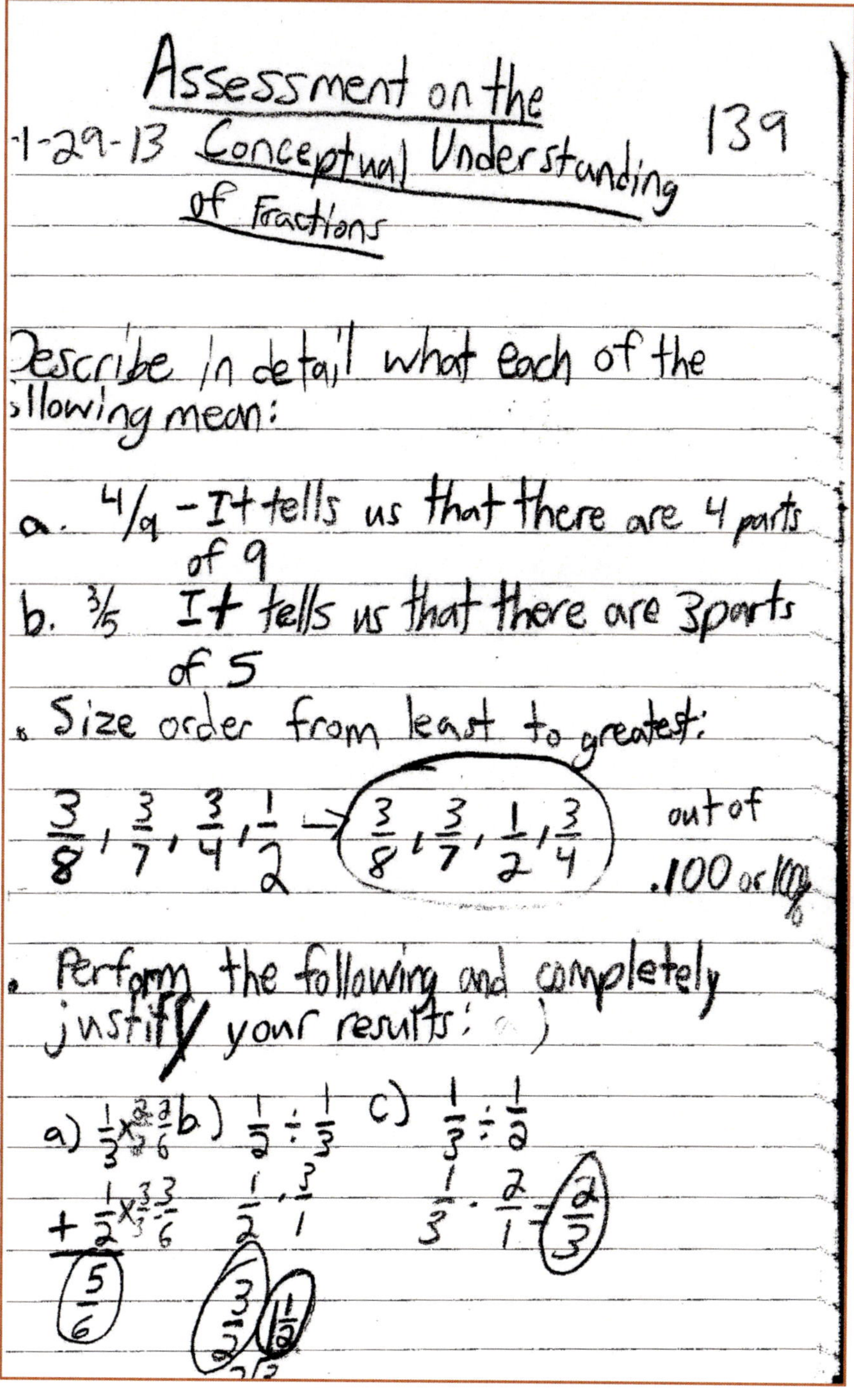

Figure 4.16: A fourth-grade learner's conceptual understanding of fractions

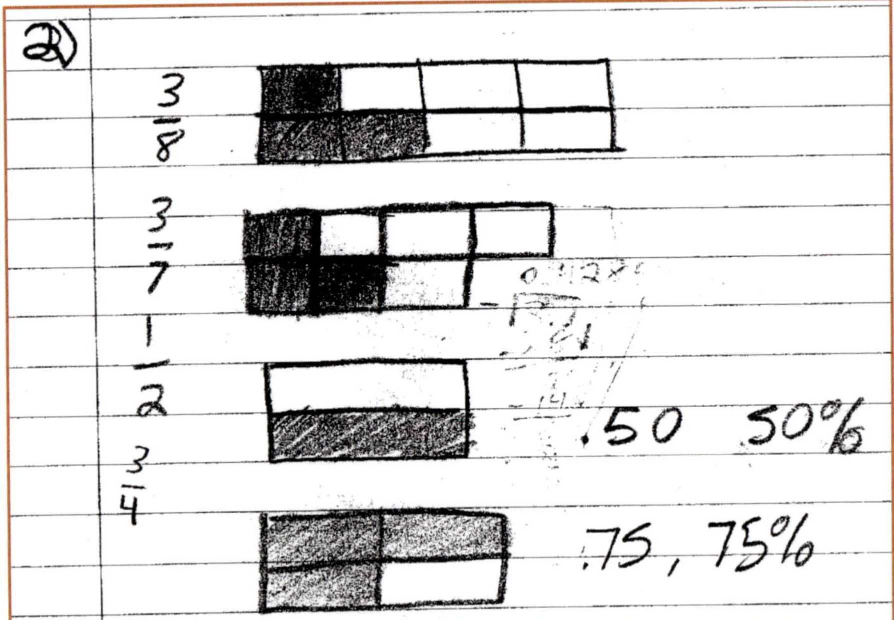

Figure 4.16: A fourth-grade learner's conceptual understanding of fractions (continued)

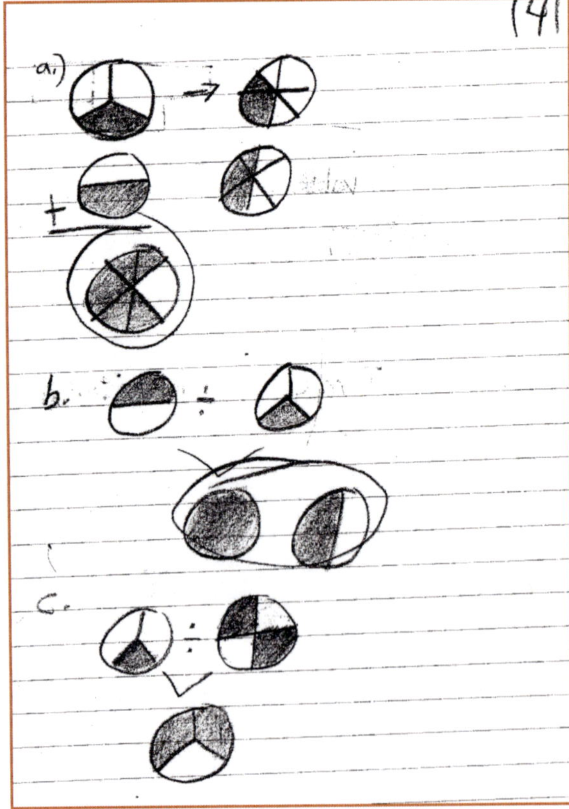

will have 2 one-halves. But I don't have three 1/3's, I have one 1/3. Therefore, I must analyze the 2 one-halves into 3 equal-sized parts, and the result will be 2/3. Thus, $1/3 \div 1/2 = 2/3$. Therefore, two-thirds of $1/2$ are in 1/3.

In Figure 4.17, an adult learner elaborates on how the cognitive function forming proportional quantitative relationships provides both a logical framework for adding and subtracting fractions and the means to construct deep conceptual understanding of the operations. The most fundamental requirement for comparing, size-ordering, adding or subtracting fractions is that all fractions involved are originating from analyzing and re-analyzing the same whole or entire quantity.

A common multiple of the denominators of the fractions involved informs the learner of the number of equal-sized parts the whole or complete quantity can be re-analyzed into to bring the fractions into a system of proportional quantitative relationships. Within such system of quantitative proportionality, the fractions essentially are "speaking the same language" or are communicating from the same logical perspective.

The RMT model teaches learners to apply their cognitive functions to construct their conceptual understanding to adding, subtracting, multiplying, and dividing fractions without resorting to meaningless algorithms. In fact, learners can use their analytical-critical thinking to derive each algorithm. Thus, through RMT, learners' conceptual understanding drives their procedural fluency and computational proficiency. Here, four of the five strands of mathematical proficiency (NRC, 2001) are visible—conceptual understanding, strategic competence, adaptive reasoning, and procedural fluency. The fifth strand, productive disposition, is emerging intrinsically within the learner.

Forming a Functional Relationship between Two Variables

Building the Cognitive Conceptual Structure

a. Size of the Circle and Content in the Circle

The RMT process transforms the pattern shown in Exhibit #25 into a mathematical event. This pattern appears on the cover page of the FIE instrument Numerical Progressions. Phenomena involving change require some conservation of constancy to detect, perceive, and define the change.

An adult learner (see Figure 4.18) applies the cognitive functions conserving constancy, comparing, labeling, forming relationships, and providing mathematical logical evidence to identify and define the variables in the pattern shown in Exhibit #25. The systemic nature of mathematics concepts is evident here, as the foundational structure for algebra is constructed. In RMT, the concept of variable embodies the core mathematics concepts of quantity and relationships. Figure 4.19 reveals that the concepts of quantity and variable become systemic elements of the broader concept, functional relationship.

Notice how the leaner moves to a high level of abstract representational-relational thinking by encoding the meaning

44

A Model for Fractions (Addition + Subtraction)
Using the cognitive function of Forming proportional quantitative relationships I am able to add and subtract fractions.

Given:
½ + ⅔

i) Find least common multiple:

For ½ I find multiples of 2
So:
2, 4, 6, 8

For ⅔ I find multiples of 3
So:
3, 6, 9, 12

2) ½ · 3/3 = 3/6

⅔ · 2/2 = 4/6

3/6 + 4/6 = 7/6 = 1⅙

Currently these two fraction are not in a proportional quantitative relationship. But I can form this relationship by changing the denominator into a common multiple. Or Equal sized parts.
I look at the multiples of the denominators and select the lowest common multiple, which in this case is 6

My new fractions now are in a proportional quantitative relationship. So I can now add these fractions.

Figure 4.17: An adult learner applies forming proportional quantitative relationships to construct understanding of adding and subtracting fractions

of each of the two variables, the size of the circle and the content in the circle, expressing the functional relationship between the two variables, and by using symbolic expression. In addition, observe the difference between the cognitive functions encoding and labeling. Encoding is an arbitrary process, where one is putting meaning into a code or symbol of choice. Labeling is giving something a name based on its properties or its critical attributes. The symbol x does not make something a variable and it certainly does not make it an independent variable. To some extent, assigning the symbol x to an independent variable and the symbol y to a dependent variable is a matter of convention in the mathematics community.

Proximity of the Dots and Overlapping of the Figures

Again, we utilize the feature of the RMT model that employs cognitive tasks as general psychological tools to build cognitive functions while concomitantly transposing the tasks into mathematical activity for constructing mathematical knowledge. The nature of some of the tasks in the FIE instruments Organization of Dots, Orientation in Space, and Numerical Progressions lend themselves to this dual role. This feature facilitates our insistence that learners' cognitive functions must be developed during the mathematics learning process.

Commencing with the Organization of Dots instrument, we use tasks to further

Exhibit #25

establish the cognitive conceptual structure for forming a functional relationship between two variables. The tasks shown in Exhibit #26 provide the structure for developing the concept, the quantity of the proximity of the dots. We are presented with a set of models—a square and two right isosceles triangles—and a row containing frames of clouds of unorganized dots. Two questions are raised: What is conserving constancy as we move from left to right? What is changing? By dynamically comparing what is in each frame and conserving constancy, we can determine precisely what is changing. The answers are the number of dots and the closeness of the dots, respectively. Closeness means proximity. Therefore, the amount or quantity of the proximity of the dots is increasing as we move from left to right.

An adult learner uses the tasks shown in Exhibits #27 and #28 to construct a functional relationship between two variables. First, the learner establishes that the proximity of the dots is a variable as

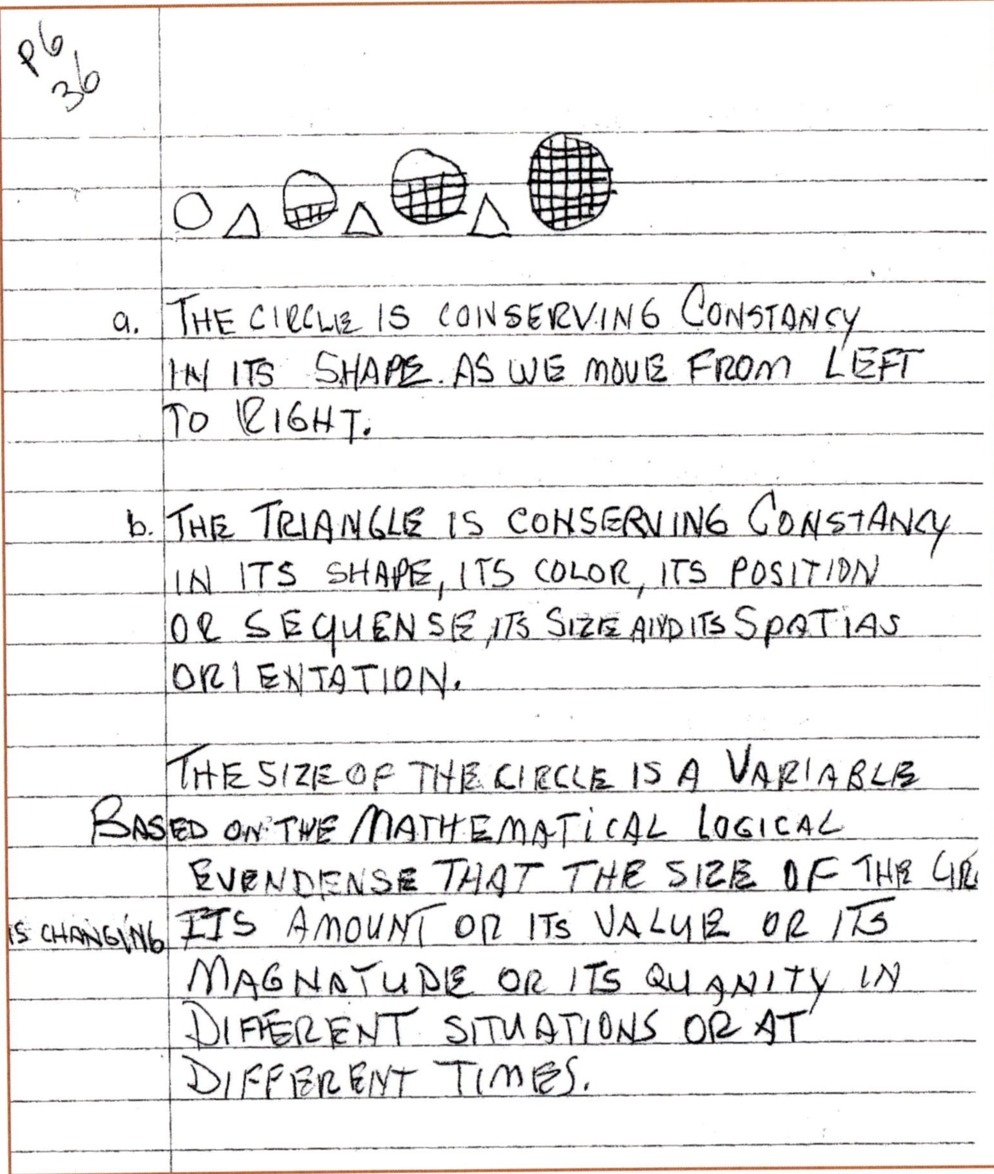

a. THE CIRCLE IS CONSERVING CONSTANCY IN ITS SHAPE. AS WE MOVE FROM LEFT TO RIGHT.

b. THE TRIANGLE IS CONSERVING CONSTANCY IN ITS SHAPE, ITS COLOR, ITS POSITION OR SEQUENSE, ITS SIZE AND ITS SPATIAS ORIENTATION.

THE SIZE OF THE CIRCLE IS A VARIABLE BASED ON THE MATHEMATICAL LOGICAL EVENDENSE THAT THE SIZE OF THE CIR IS CHANGING ITS AMOUNT OR ITS VALUE OR ITS MAGNATUDE OR ITS QUANITY IN DIFFERENT SITUATIONS OR AT DIFFERENT TIMES.

Figure 4.18: An adult learner applies cognitive functions to define the variables in the pattern

FORMING A FUNCTIONAL RELATIONSHIP
BETWEEN TWO VARIABLES

PG 37

1. DOES ONE VARIABLE DEPEND ON THE OTHER?
YES - THE CONTENT IN THE CIRCLE IS
DEPENDENT ON THE SIZE OF
THE CIRCLE.
THEREFORE, THE SIZE OF THE CIRCLE
IS THE INDEPENDANT VARIABLE AND
THE CONTENT OF THE CIRCLE IS THE
DEPENDENT VARIABLE.

2. IS THIS A CAUSE-EFFECT RELATIONSHIP?
YES
THE SIZE OF THE CIRCLE IS THE CAUSE
VARIABLE.
AND THE
CONTENT IN THE CIRCLE IS THE EFFECT
VARIABLE

3. IS THIS AN IN-PUT OUT-PUT RELATION?
YES
THE SIZE OF THE CIRCLE IS THE IN-PUT VARIABLE
AND THE
CONTENT IN THE CIRCLE IS THE OUT-PUT VARIABLE

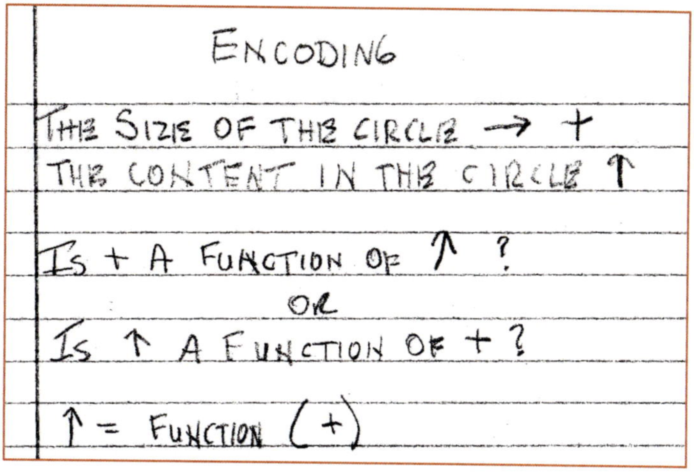

he moves down the page

Figure 4.19: An adult learner formulates structure for forming a functional relationship between two variables

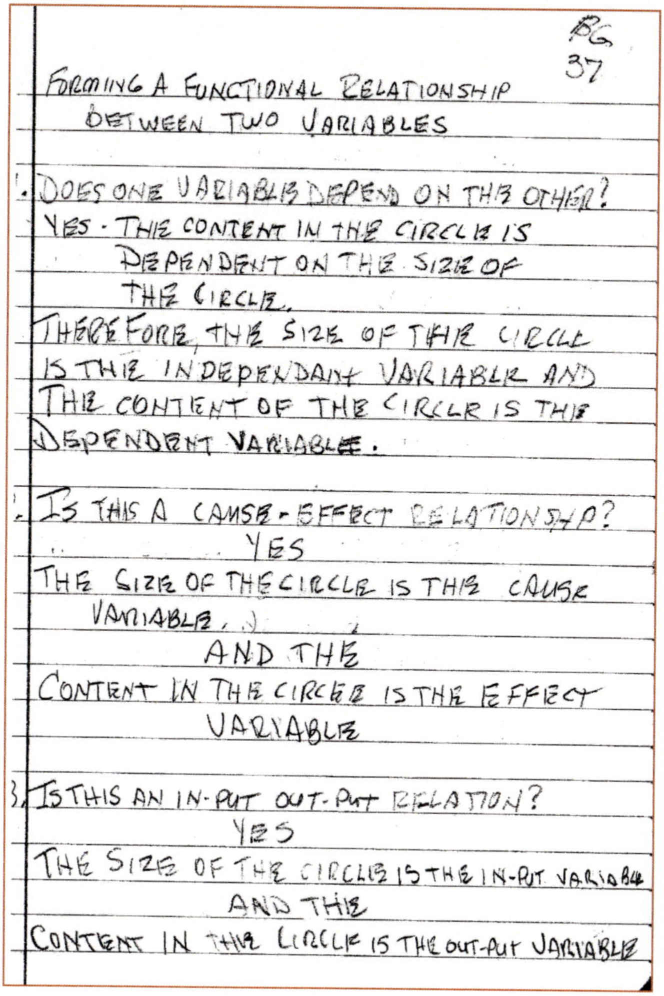

ENCODING

THE SIZE OF THE CIRCLE → +
THE CONTENT IN THE CIRCLE ↑

IS + A FUNCTION OF ↑ ?
OR
IS ↑ A FUNCTION OF + ?

↑ = FUNCTION (+)

in Exhibit #27. Next, the learner establishes that the overlapping of the figures is a variable as he moves down the page shown in Exhibit #28. Using these same sets of stimuli, he expresses how the proximity of the dots brings about the overlapping of the figures.

Orientation of the Boy and His Relationships with Objects

Turning to the Orientation in Space I instrument in FIE, we use tasks (see Exhibit #9) to further enrich the cognitive conceptual structure for forming a functional relationship between two variables. As the quantity or amount or value or magnitude of the spatial orientation of the boy changes by v degrees clockwise, the relationship between the boy and each of the four objects changes its amount by v degrees counterclockwise. For example, when the boy rotates 90⁰ clockwise from position A to position B, his relationship to the house changes by 90⁰ counterclockwise, front to left. In another example, when the boy changes his spatial rotation by 270⁰ clockwise, from position A to position D, the relationship between the boy and the house changes by 270⁰ counterclockwise from his front to his right. Performing these tasks with the full use of mathematical language, the general cognitive function **hypothetical thinking** is transformed into the mathematically specific cognitive function, **mathematical inferential-hypothetical thinking.**

Forming a Functional Relationship between Two Variables

In Figure 4.21, an adult identifies three mathematically specific psychological tools—a formula or equation, mathematical language, and a table—that are used in RMT to form a functional relationship between two variables. These are no longer symbolic representations to learners who have appropriated each and internalized its unique structure.

In Figure 4.21, this learner illustrates that the structure of the mathematical relationship always conserves constancy between the total quantity, amount, or magnitude on the left side of the equal sign with the total quantity, amount, or magnitude on the right side. This structure of the tool helps the learner to logically transform any linear algebraic equation in two variables from the standard

Exhibit #26

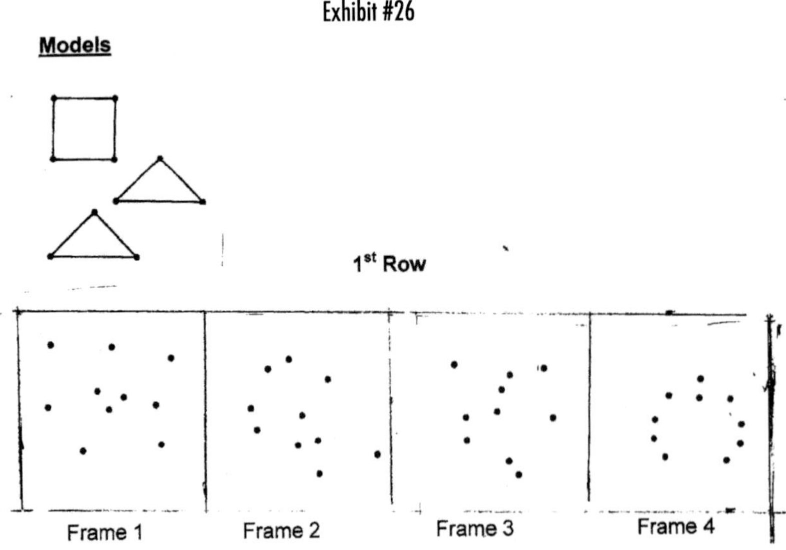

132

Exhibit #27

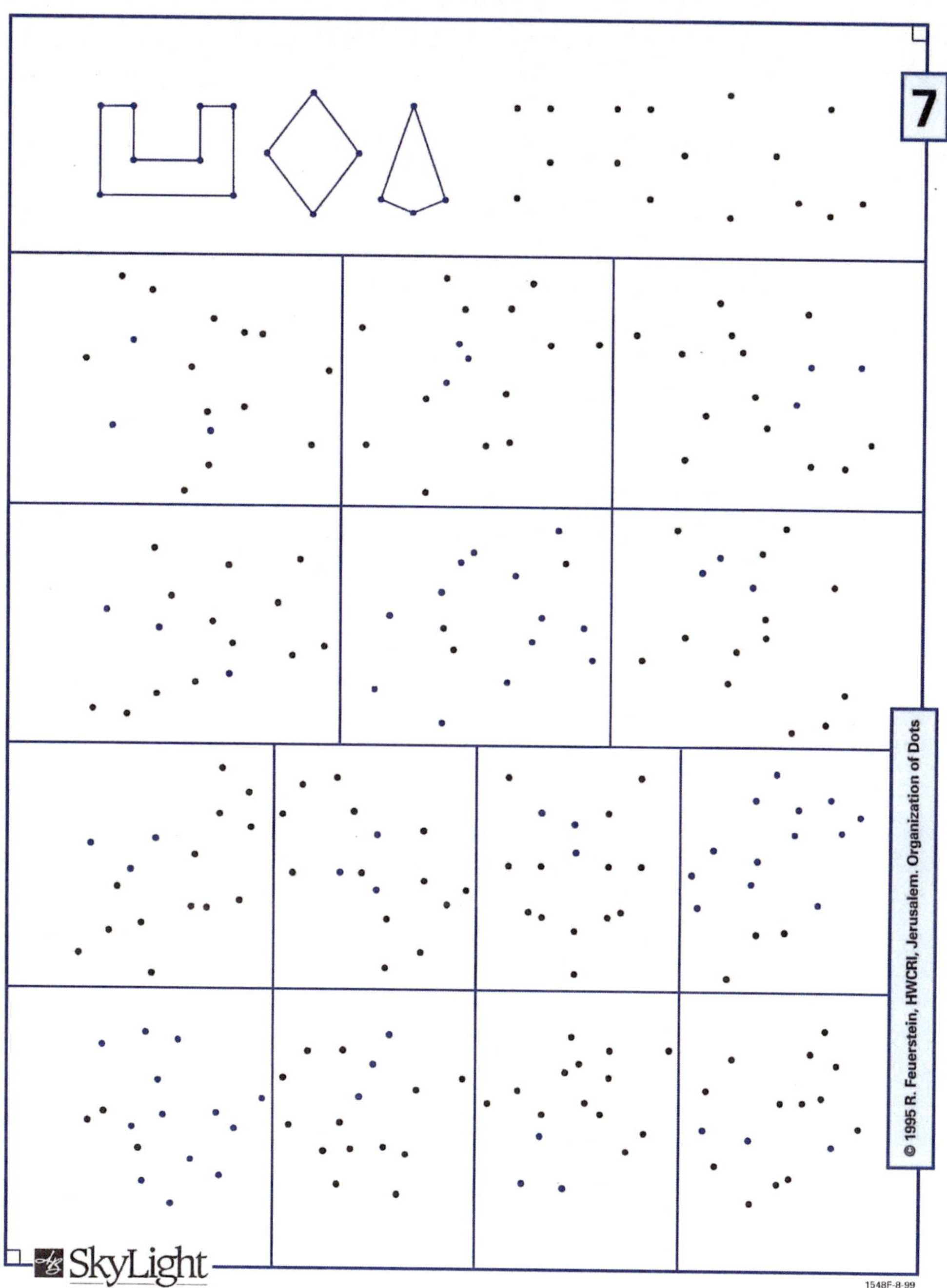

SkyLight
PROFESSIONAL DEVELOPMENT

1548F-8-99

Exhibit #28

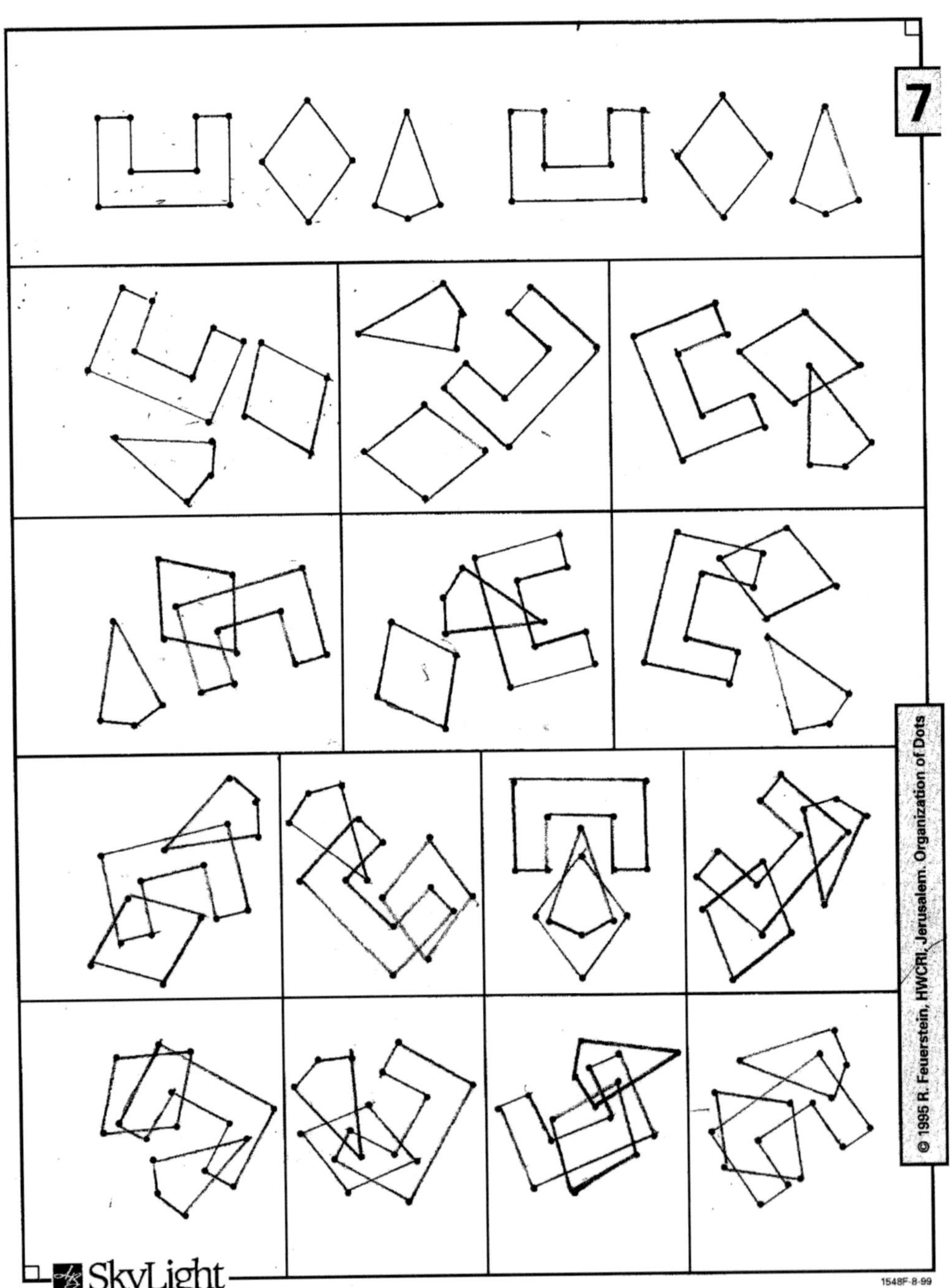

1548F-8-99

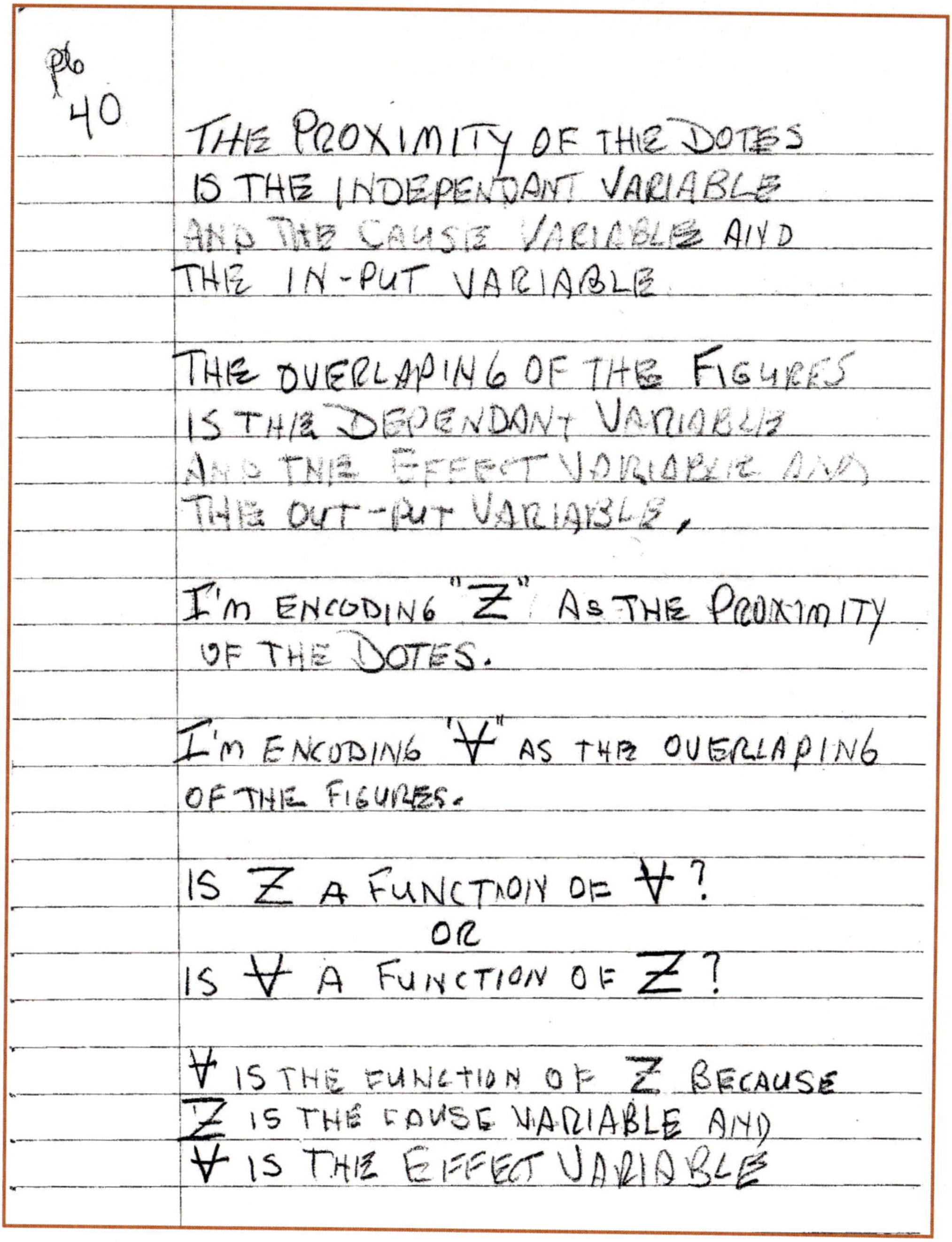

pg 40

THE PROXIMITY OF THE DOTES
IS THE INDEPENDANT VARIABLE
AND THE CAUSE VARIABLE AND
THE IN-PUT VARIABLE.

THE OVERLAPING OF THE FIGURES
IS THE DEPENDANT VARIABLE
AND THE EFFECT VARIABLE AND
THE OUT-PUT VARIABLE,

I'M ENCODING "Z" AS THE PROXIMITY
OF THE DOTES.

I'M ENCODING "∀" AS THE OVERLAPING
OF THE FIGURES.

IS Z A FUNCTION OF ∀?
OR
IS ∀ A FUNCTION OF Z?

∀ IS THE FUNCTION OF Z BECAUSE
Z IS THE CAUSE VARIABLE AND
∀ IS THE EFFECT VARIABLE

Figure 4.20: Adult learner formulates structure for forming a functional relationship between two variables using a figural modality

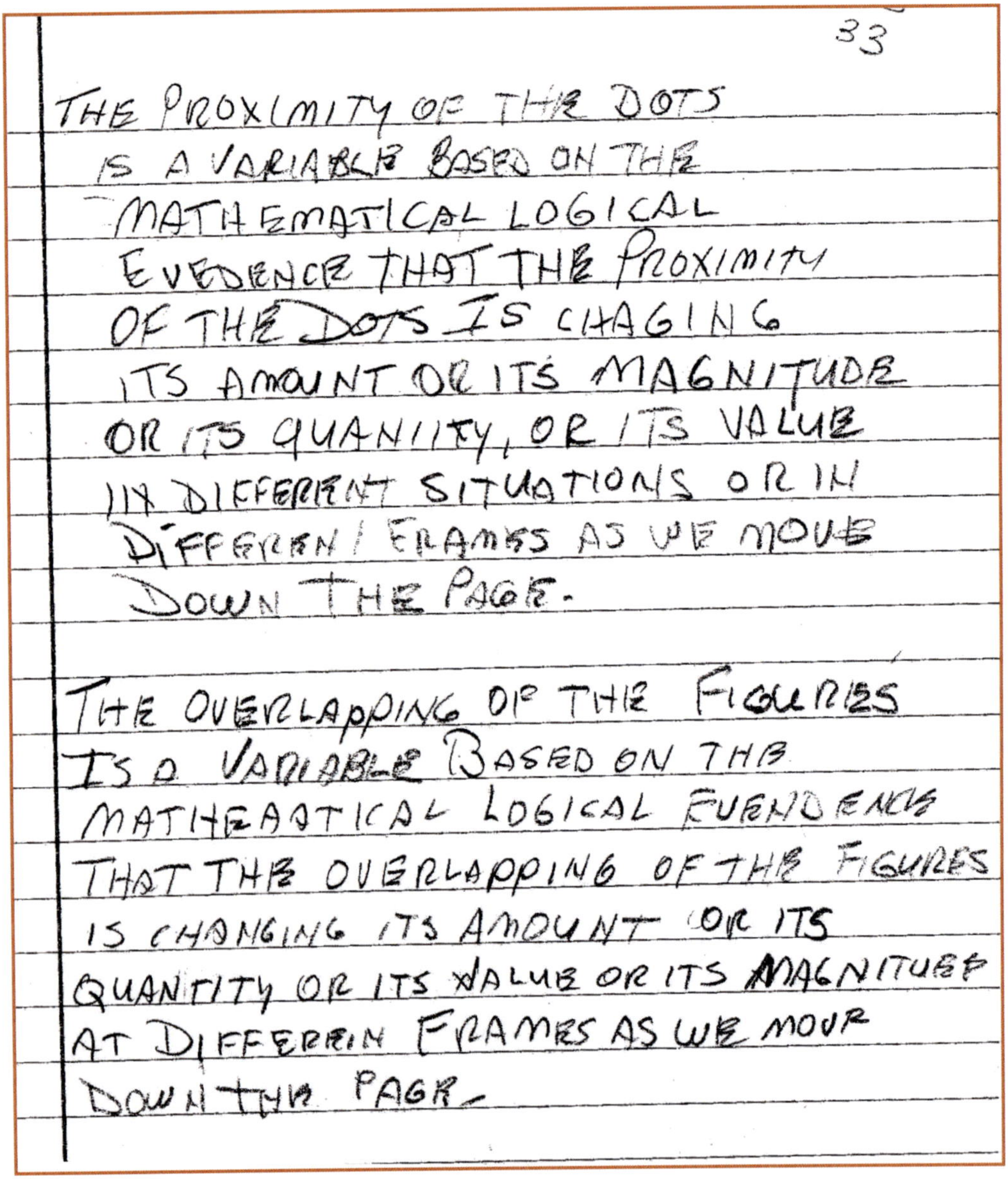

33

THE PROXIMITY OF THE DOTS IS A VARIABLE BASED ON THE MATHEMATICAL LOGICAL EVEDENCE THAT THE PROXIMITY OF THE DOTS IS CHAGING ITS AMOUNT OR ITS MAGNITUDE OR ITS QUANIITY, OR ITS VALUE IN DIFFERENT SITUATIONS OR IN DIFFERENT FRAMES AS WE MOVE DOWN THE PAGE.

THE OVERLAPPING OF THE FIGURES IS A VARIABLE BASED ON THE MATHEMATICAL LOGICAL EVENDENCE THAT THE OVERLAPPING OF THE FIGURES IS CHANGING ITS AMOUNT OR ITS QUANTITY OR ITS VALUE OR ITS MAGNITUBE AT DIFFEREIN FRAMES AS WE MOVR DOWN THE PAGR

Figure 4.20: Adult learner formulates structure for forming a functional relationship between two variables using a figural modality (continued)

form into the slope-intercept form. The slope-intercept form, generally expressed as y = mx + b, more readily facilitates the **forming a functional relationship between the two variables.**

In RMT, mathematical language must be used by the learner with the other mathematically specific psychological tools to form, apply, and construct deep understanding of the functional relationship between the two variables. For example, for the equation y = 3x + 5, a learner speaks openly or to himself or herself: "For every amount, quantity, magnitude or value of the independent variable or the cause variable or the input variable, x, I multiply it by 3 to get a product, and to that product I add the quantity 5 to get the corresponding amount, quantity, magnitude, or value of the dependent variable or the effect variable or the output variable, y."

In Figure 4.22, a ninth-grade learner solves a system of linear equations by using equations, tables, and an x-y coordinate plane as mathematically specific psychological tools to form functional relationships between the two variables.

Forming a Unit Functional Relationship (Slope)

In the RMT paradigm, the concept of slope is also the mathematically specific cognitive function, **forming a unit**

functional relationship. It is also a part of the systemic conceptual structure of the big mathematical idea, forming proportional quantitative relationships. When forming a unit functional relationship, one is stating the change in the quantity of the dependent variable or the effect variable or the output variable that is produced by a unit change in the quantity of the independent variable or cause variable or input variable. The RMT practice of mediating learners to actively construct and apply cognitive functions during the mathematics learning process takes place through our mathematical learning activity (MLA).

The following three criteria must be met for an activity to qualify as MLA (Kinard and Kozulin 2008, pp. 2-3):

1. The activity should aim at creating a structural change in the learner's understanding of mathematical knowledge. A mere accumulation of information or skills does not qualify as a structural change. By structural change, we mean a qualitative change in the learner's level of mathematical comprehension. Such a change is characterized by systemic organization, self-regulation, and transformation. A new level of comprehension should have a systemic unity so that all the elements become involved in it. In addition, the change should, on the one hand, be strong enough to withstand the temptation to solve

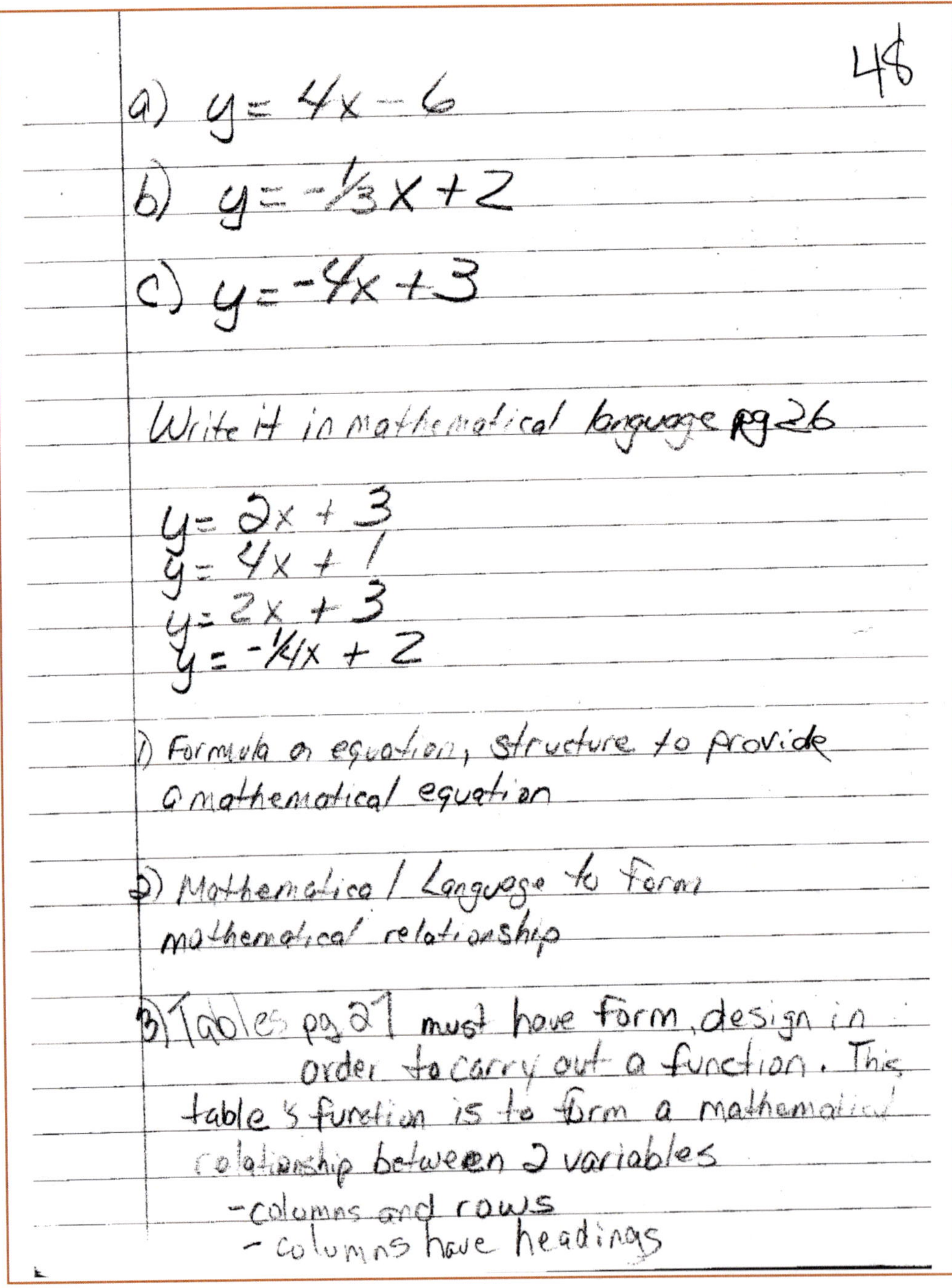

Figure 4.21: Adult learner identifies three mathematical psychological tools to form a functional relationship between two variables

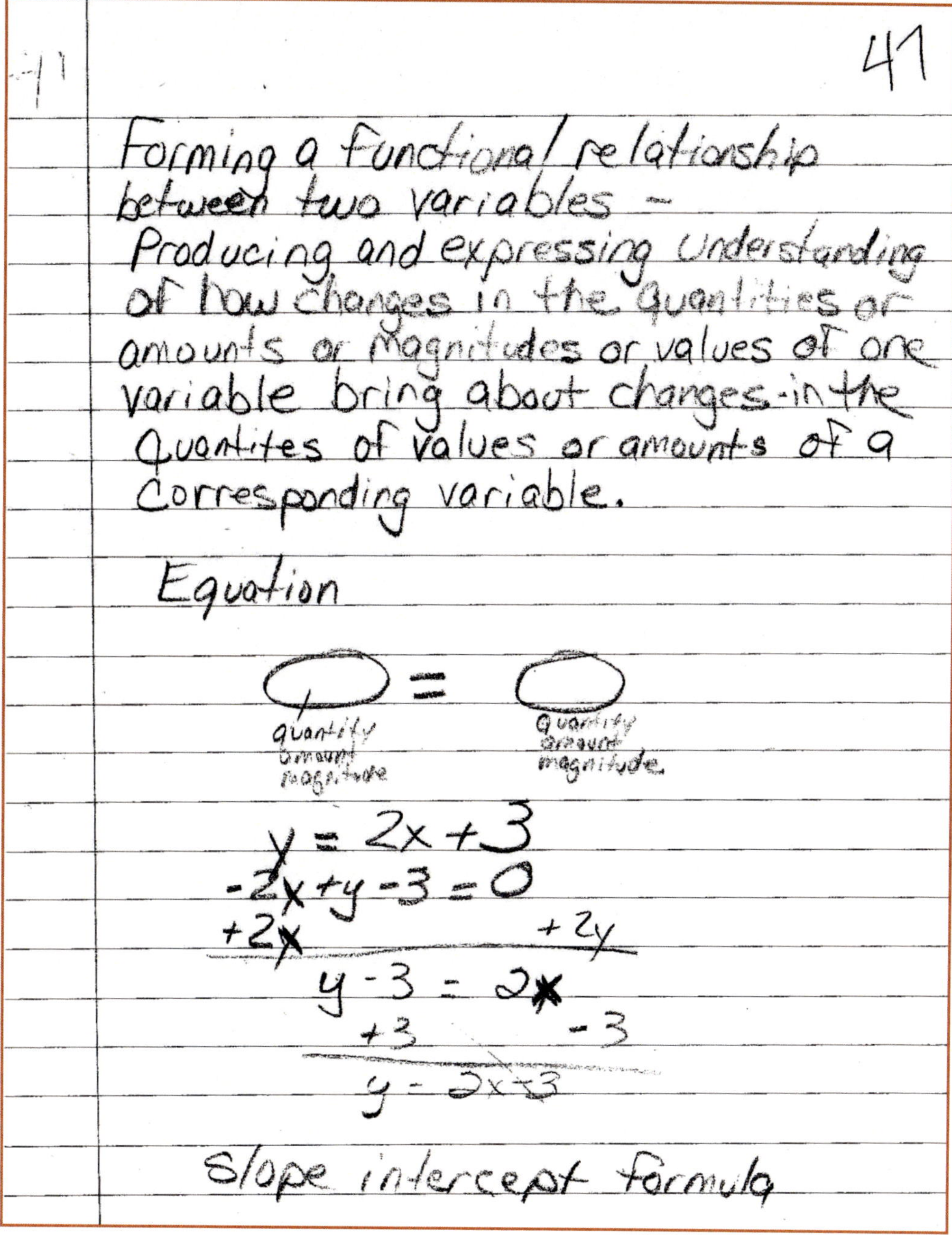

Figure 4.21: Adult learner identifies three mathematical psychological tools to form a functional relationship between two variables (continued)

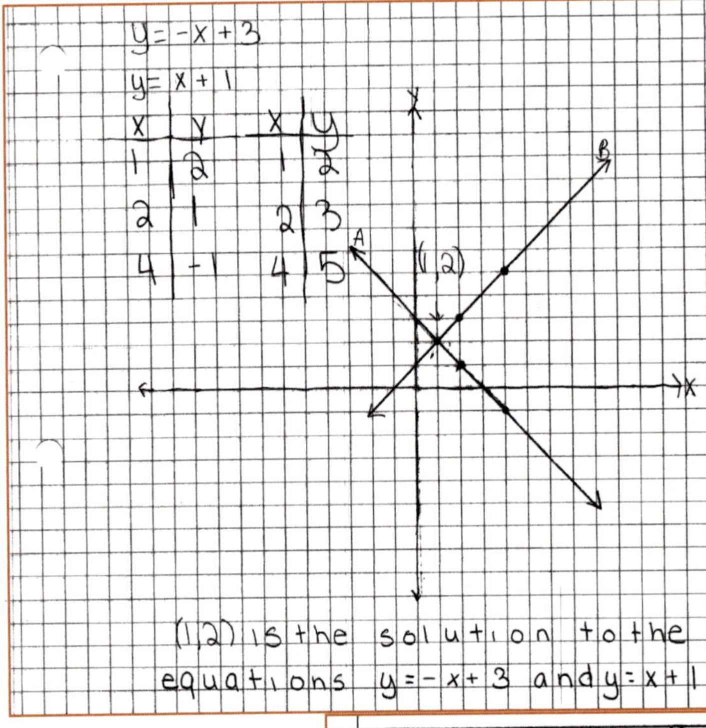

Figure 4.22: A ninth-grade learner uses mathematical psychological tools and cognitive functions to solve a system of linear equations

The cognative functions I used to find the solution to this equation and graph are forming relationships, visualizing and decomposing. We are forming relationships by calculating the sum value when we substitute x and y. We are visualizing by using the graph as a tool to visually display the two equations in a linear format. Lastly we are decomposing because when we substitute values for x and y we are essentially breaking down the equation to see which two coordinates would be a solution to the problem. Also with the correlation of the slopes, line B is positive and line A is negative. This is forming a relationship because even though the lines have opposing slopes they still intersect at (1,2).

novel tasks by reverting to previous, less advanced levels of reasoning and, on the other hand, open enough for further transformation. This new level of understanding is produced by the learner's use of cognitive functions and psychological tools to organize and form relationships between supporting conceptual elements;

2. The classroom activity must aim toward and, therefore, be a part of a process for constructing a "scientific" mathematics concept, characterized by its theoretical, generative, and systematic nature. The genetic analysis of mathematics concepts, their derivation from measurement, and representation by schematic modeling, differs substantively from both historical and current U.S. reform efforts. An example of a theoretical concept is the following: When the mediator asks learners why a cork floats in a tub of water and a nail sinks, learners respond that the nail is long and thin and the cork is more round. To counter this misconception, the mediator places a wooden matchstick and a steel ball bearing in the water and the learners observe that the long thin matchstick floats and the steel ball bearing sinks. Clearly, the theoretical concept of density must be mediated to the learners. This concept is theoretical because it cannot be grasped empirically (observed in the physical world), but it requires a theoretical mode of thinking for its

appropriation; and

3. The learning activity must introduce learners to the language and rules of the mathematics culture. Whatever the learners' native language, family customs, or everyday experiences, the collaboration in the mathematics classroom should provide them with an opportunity to be introduced to a culture new to all of them and one that is becoming common to all of them—the culture of mathematical thought. This culture has its own language, rules and customs, history, and challenges.

The presence of all three of these criteria of MLA is evident in the reflections on slope written by a ninth-grade learner (see Figure 4.23) and an adult learner (see Figure 4.24).

In traditional math classes, students are taught that the slope is the rise over the run. In this work, the ninth-grade RMT learner and the adult RMT learner are demonstrating a structural shift in their thinking and comprehension. The change is from a more mechanical approach, rise over run, to a more conceptual dynamic understanding, rate of change in the values of the two variables. Both learners are using mathematical language to drive and articulate their thinking and understanding.

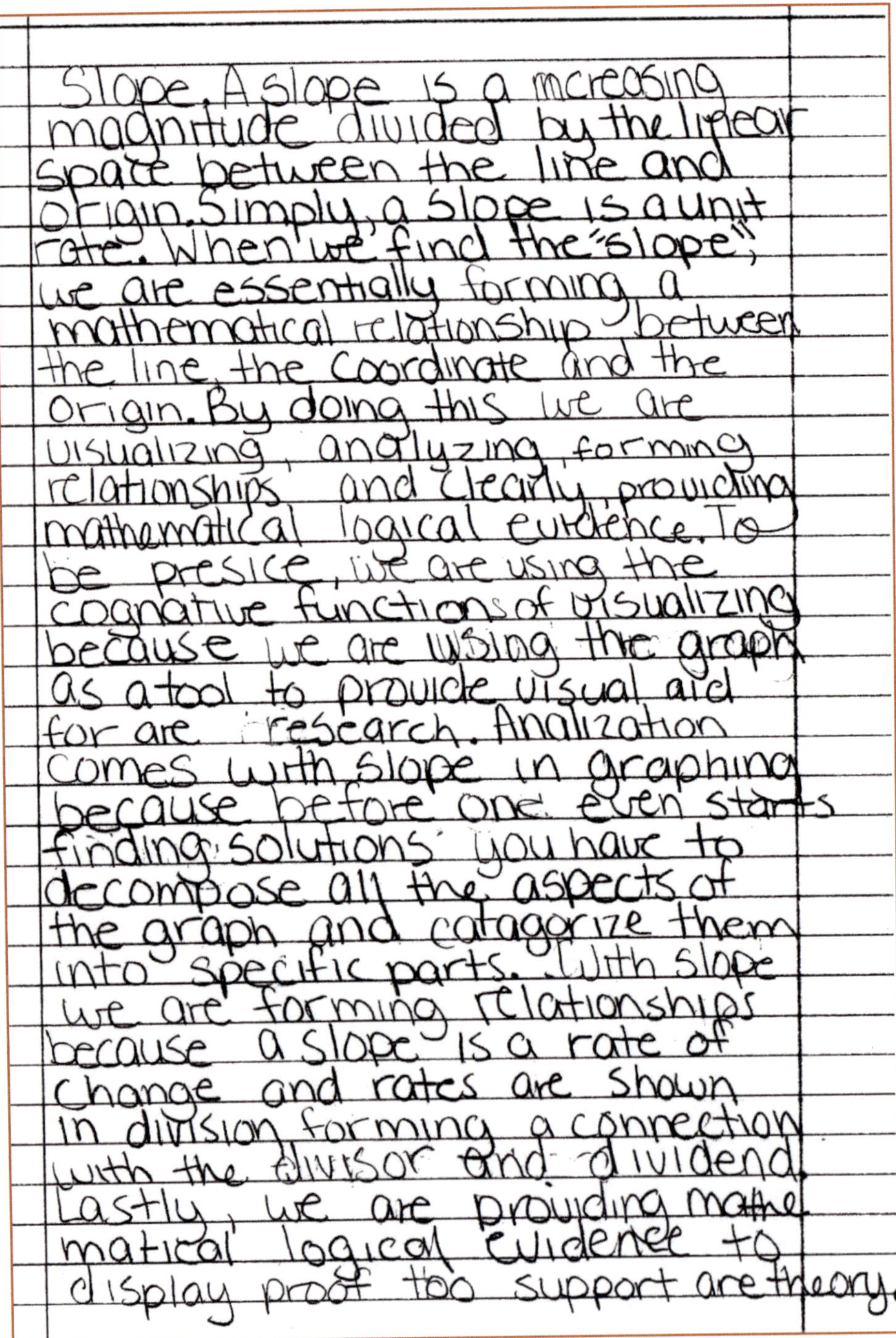

Figure 4.23: A ninth-grade learner's evidence of undergoing a structural change in her understanding of slope

Slope

the measure of the steepness or slant of a quantity of linear space (line)

Slope = Rise (change in y) / Run (change in x)

positive slope slants upward
negative slope slants down

positive slope   negative slope

zero slope | undefined slope

↕ Rise is the amount of vertical distance between points
  – focus on y axis

↔ Run is the amount of horizontal distance between points
  – focus on x axis

$m = \dfrac{y_2 - y_1}{x_2 - x_1}$

Slope-intercept $y = \boxed{mx + b}$ including sign
                        slope   is the y intercept

Ex. $y = 2x + 5$ $y = -3x - 2$
    slope = 2   slope = -3
    y intercept = 5   y intercept = -2
        y intercept always (0, b) because it is on the y-axis

(cont) pg 48
– Columns form a set (independent variables)
  set
4) x-y coordinate plane

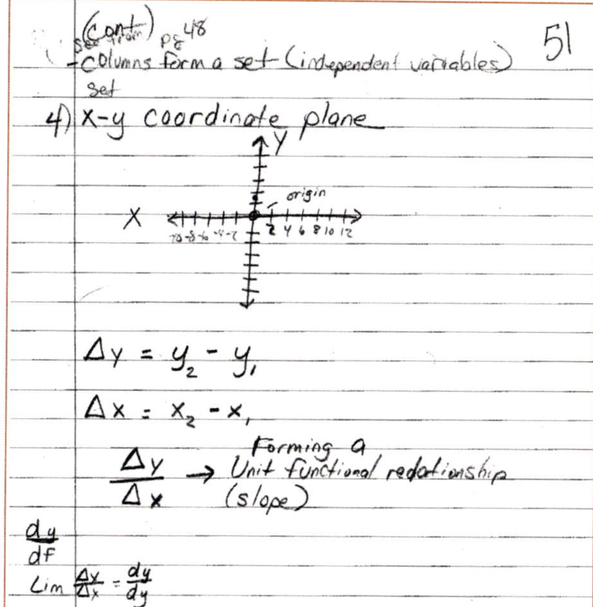

$\Delta y = y_2 - y_1$

$\Delta x = x_2 - x_1$

$\dfrac{\Delta y}{\Delta x}$ → Forming a Unit functional relationship (slope)

$\dfrac{dy}{df}$

$\lim \dfrac{\Delta y}{\Delta x} = \dfrac{dy}{dy}$

Figure 4.24: An adult learner's evidence of undergoing a structural change in her understanding of slope

Understanding the Formal Mathematical Definitions of Relations and Functions

According to the RMT model, an ordered pair is a quantity or value of the independent variable and the corresponding quantity or value of the dependent variable that provides a solution to the functional relationship between the two variables. The entire set of ordered pairs that satisfy the functional relationship is a quantitative or numerical expression of this functional relationship. All quantities or values of the independent variable in this set of ordered pairs belong to the domain of this functional relationship. All quantities or values of the dependent variable of this set of ordered pairs belong to the range of this functional relationship. This knowledge that RMT learners have constructed equips them to readily understand the formal mathematical definitions for relation and function given below.

Formal Definitions

A relation is a set of ordered pairs. The set of the first entries of the ordered pairs is the domain (D) of the relation. The set of second entries is the range (R) of the relation. The points in the coordinate plane that correspond to the ordered pairs of the relation form the graph of the relation.

A function is a relation that assigns a single element of R to each element of D.

Understanding and Applying Physics Concepts through RMT Model

Scientific ideas are generated through observations of nature, thinking, experimentation and validation. However, every major physics concept has an underlying mathematical structure, which is essentially theoretical. For example, force is described as a push or pull on an object. Descriptions of numerous observations and experiments in pushing and pulling objects can be produced, but without the mathematical structure $F = ma$ and its underlying conceptual meaning, coherent thinking and precise quantitative understanding and applications of this idea would not be possible. In a previous publication, Kinard and Kozulin (2008, pages 182–188), reported data from a 76-hour RMT intervention as part of a larger six-month training that aimed at preparing inner-city residents with environmental remediation and construction skills. Learners, who were African-American males 16–24 years of age, engaged in constructing mathematical concepts and applying them to understanding physics concepts, such as velocity, acceleration, momentum, force, relativity, etc. During the last phase of the intervention, these learners were mediated to perform selected tasks of advanced psychological tools contained in the FIE instruments Orientation in Space I and Numerical Progressions. Mediation to perform these tasks promoted the learners' development and application of cognitive functions needed to identify and define motion, change, change within change, and dynamic relationships.

Learners were mediated to begin defining the problem given in Exhibit #29. Analyzing the complete field of data into its major parts was essential in determining and identifying the known. Following are the results of this analysis:

1. A compass rose at the top middle of the page with units of quantities of angular movement and directions of movement.
2. Six fill-in-the-blank tasks on circular phenomena and angular movement.

The tasks require the construction of multiple logical pathways that connect the known to the unknown. This emphasizes that the RMT approach to problem solving does not focus primarily on getting an answer to a problem, but on designing and evaluating problem-solving strategies that can create new insights. This set of tasks provides learners practice in applying cognitive functions to angular measure and dynamically changing units of measure in the context of change and abstract relationships.

The psychological tools of the Numerical Progressions instrument lend themselves to mediating learners to perform tasks that are characteristic of mathematical thinking. These tasks engage learners to search for and define stable relationships between specific experiences; formulate the rules and laws that govern a succession of events; and explore the order and rhythm of relationships and their appearance in cycles of progressions (Feuerstein and Hoffman, 1995, p. 2). Samples of these tasks are given in Exhibit #30. The first picture shows that the boy ascends or moves up the stairs by taking one step forward. Thus, his motion of ascending has two simultaneous components, a forward or horizontal movement by one step and an upward or vertical movement by one step. The second picture shows the reverse motion of the boy descending by moving forward or horizontally one step and downward or vertically by one step.

The numerical progression in item 1 shows a sequential descending numerical movement, starting at the progression's first member, 11, and forming a relationship of -1 between adjacent members. Thus learners, through their inductive thinking, can formulate that the relationship between successive or adjacent members of the progression defines both the magnitude and direction of movement in the progression. This descending progression then has a formula with the structure of -1 above and in the middle of two blank spaces. The -1

Exhibit #29

11

A *turn* is 1/4 circle.
One turn includes 90º.

A *half turn* is 1/8 circle
or one half of a turn.
One half turn includes
45º.

For a person who
is turning around.

____

right turn

left turn

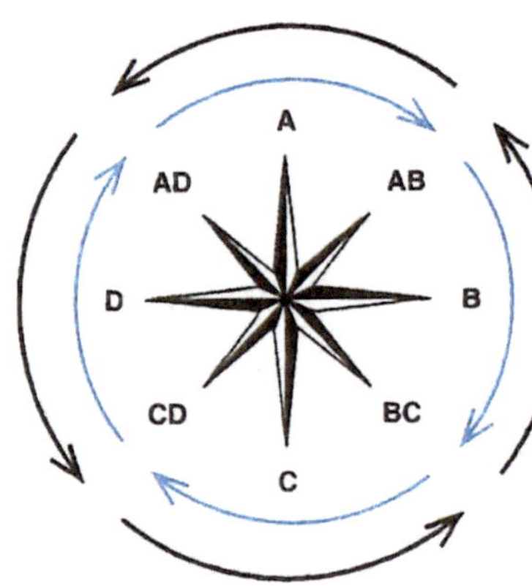

1. In order to complete a full circle, you do ________ turns

   or ________ half turns

   or 2 turns + ________ half turns

   or ________ turns+ ________ half turns

   or ________________ + ________________

2. A person is standing in the middle of the circle.
   In order to turn from A to BC, he or she must do
   one turn + ________________ to the right,
   or ________________________ to the left.

3. In order to turn from AD to BC, he or she must do
   ________________________ to the right,
   or ________________________ to the left.

4. In order to turn from D to BC, he or she must do
   ____

   or ________________________________

5. In order to turn from C to ________ , he or she must do
   two turns and three half turns rightward,
   or ________________________________

6. In order to turn from ________ to AB, he or she must do
   ________________________________ clockwise

   or one turn and 4 half turns counterclockwise.

> For the person standing in the middle of the circle:
> turning clockwise/counterclockwise is a movement rightward;
> turning clockwise/counterclockwise is a movement leftward.

is in the circle, forming the quantitative relationship between the blank space on the left, generally represents the value of any member, and the next blank space representing the value of the adjacent member. This latter value will be 1 less than the value of the previous member. This formula can now be used to perform deductive thinking to expand or continue the progression. The formula expresses a decreasing quantitative rate of change of -1 between consecutive members.

Further analysis reveals a second subtle element of change as the progression moves from left to right, the position of the members or the time of their ap-

pearance. Thus, there are two variables, time and the member of the progression. And they are forming a functional relationship with each other. If we encode time as x and member of the progression as y, y = -x + 11. This equation is in the generalized slope intercept form of a linear equation, y = mx + b. The invisible -1 that is the coefficient for the variable time in this equation is the slope. The cognitive function **forming a unit functional relationship** in this case is the following: the member of the progression decreases in value by 1, for every unit increase in the value of time. This is a decreasing or descending rate. Movement occurs in the progression in

Exhibit #30

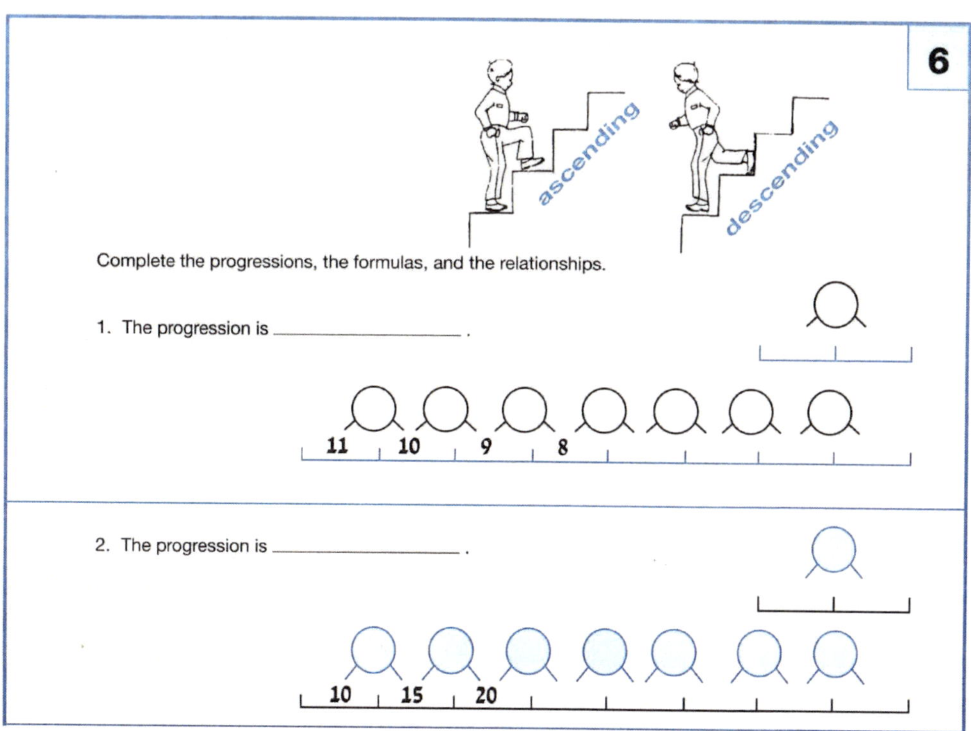

item 2 with a rate of increase of 5 between adjacent members. Learners can apply mathematical inductive thinking to the progression and generate the linear functional relationship y = 5x + 10. The quantity or the amount of the member of the progression increases or ascends by 5 for every unit increase in the amount of time.

Exhibit #31

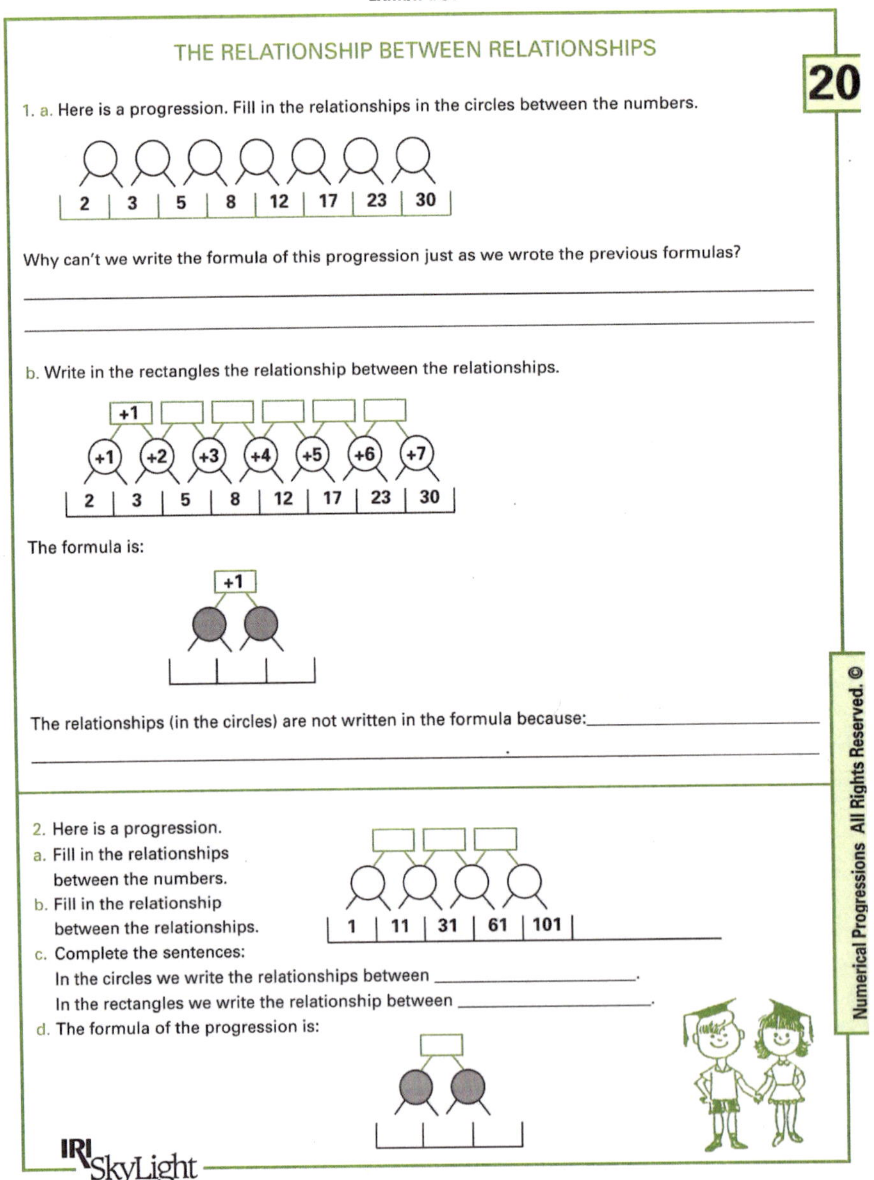

Learners are confronted with disequilibrium or dissonance when forming relationships between the members of the first progression presented in Exhibit #31. The source of this issue is the lack of a stable, unchanging relationship or set of relationships that clearly defines the quantitative movement in the progression, which can be used to formulate and predict future events. This source of difficulty can be further elaborated as change within change by analyzing item 1b. Learners must form relationships between relationships to arrive at a stable, unchanging transcendent relationship. Such relationships conserve constancy in the rate of change throughout the movement within the progression.

After mediating learners to engage in the above cognitive conceptual construction, they were given the following mathematical-scientific investigation project to work on individually and in their small collaborative teams.

Mathematical-Scientific Investigation

Consider an object with a mass of two pounds dropped from a spaceship in proximity to the surface of a planet with no atmosphere. Gravitational pull on the object causes it to fall according to the data given in the table below.

Time object has Fallen (seconds)	Distance object has Fallen (feet)
1	5
2	31
3	76
4	140
5	223
6	325
7	446

Without any further information or instructions, learners immediately began to work individually. Shortly thereafter, they collaborated. Later during mathematical-scientific discourse, one of them placed the data shown in Figure 4.25 on the board.

Vignette

Mediator: What is this you have constructed on the board?

Learners: The first progression at the bottom goes from left to right. It is the time in seconds that the object has fallen from the spaceship.
❑We are looking at everything by going from left to right.
❑The numbers in the circles show the

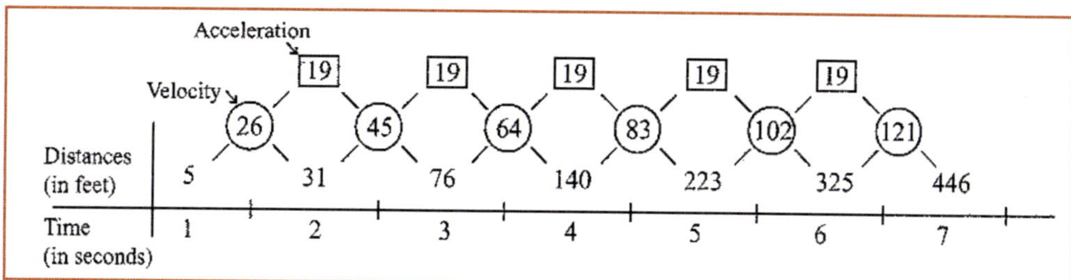

Figure 4.25: A small collaborative team of learners' work on investigation

relationships between the distances the object has fallen. The first number is twenty-six.

❑This is the number of feet the object falls from one second to two seconds. It is also the distance from five feet to thirty-one feet.

❑The number in the second circle is the distance the object falls from two seconds to three seconds. It is also the distance the object falls from thirty-one feet to seventy-six feet.

❑These relationships form a progression also.

❑Each relationship is the velocity of the object.

Mediator: Explain!

Learners: Well, the object is falling toward the surface of the planet. We learned earlier that the velocity is the distance of movement per unit of time in a certain direction. From five feet to thirty-one feet, the velocity of the object is twenty-six feet per second. From thirty-one feet to seventy-six feet, the object's velocity is forty-five feet per second.

❑So the velocity of the object is changing and it is a variable. The distance the object has fallen and the time it has fallen are also variables.

❑There is a functional relationship among the two variables, the time the object has fallen, and the distance the object has fallen.

❑There is also a functional relationship between the distance the object has fallen and the velocity of the object.

Mediator: What is occurring at the next level of your diagram?

Learners: Here we have the relationships between relationships.

❑We also can say there is a functional relationship between the time the object is falling and the object's velocity.

❑Here we are seeing how the velocity of the object is changing over time. The velocity is in feet per second. This is how the distance of the object changes with time. Now we are seeing how the velocity of the object is changing with time. This is the feet the object falls per second per second.

❑This is the acceleration of the object as it falls toward the planet.

❑The object is accelerating nineteen feet per second per second.

❑This acceleration is the change in the object's velocity per unit time.

Mediator: Can you explain this in a different way?

Learners: Yes. With each second that passes, the object increases its speed toward the planet by nineteen feet per second.

❑The acceleration of the object toward the planet is conserving constancy. It is uniform over the distance and time the object is moving.

Mediator: Where is the acceleration coming from?

Learners: It is coming from the gravitational pull the planet is exerting on the object toward its surface at nineteen feet per second per second.

Mediator: Is there any significance that the object has a mass of two pounds?

Learners: Activating my prior knowledge, we learned that force is the pull or push on an object of a certain mass to accelerate its motion or to change its velocity. This is the force it takes to move this two-pound object nineteen feet per second per second. This means that the force on the object is thirty-eight pounds per foot per second per second.

Mediator: Could you call the presentation you have constructed on the board a system of psychological tools?

Learners: Sure. We used the data given to us on the object falling from the spaceship to form these progressions, relationships, and relationships between relationships that caused us to use many cognitive functions and helped us to more deeply understand force.

❑We have been building our understanding around the formula F=ma or force equals mass times acceleration. This formula and the psychological tools on the board have helped us to integrate cognitive functions like forming proportional quantitative relationships, projecting and restructuring relationships, quantifying space and spatial relationships, quantifying time and temporal relationships.

❑As we analyzed and integrated the data about the falling object we constructed and use the tools of the progression.

❑This required us to use those cognitive functions you mentioned and others. The relationships among the time distance, velocity, acceleration, mass, and force connect with the formula F=ma. This has really helped me to develop a deeper understanding of force.

❑While we were talking, I thought about using the mathematical psychological tool of the x-y coordinate plane to graph the functional relationship between the time the object has fallen and the distance the object has fallen (see Figure 4.26) and the time the object has fallen and the velocity of the object (see Figure 4.27).

Mediator: Will you put them on the board please? [Learner draws graphs on the board]

Learners: The functional relationship between the time the object has fallen and the velocity of the object is linear. As we said before, the acceleration of the object conserves constancy.

Mediator: This is very good work!

This demonstrates the potential for inner-city learners to become brilliant mathematicians and scientists through systemic engagement in cognitive conceptual construction. ■

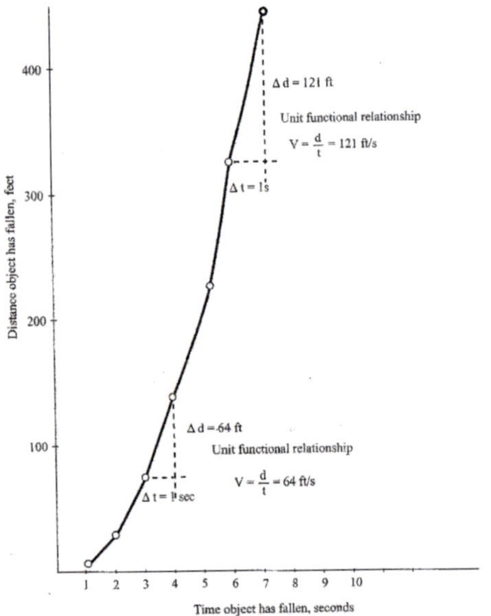

Figure 4.26: Learner's graph of the time-distance relationship the object has fallen

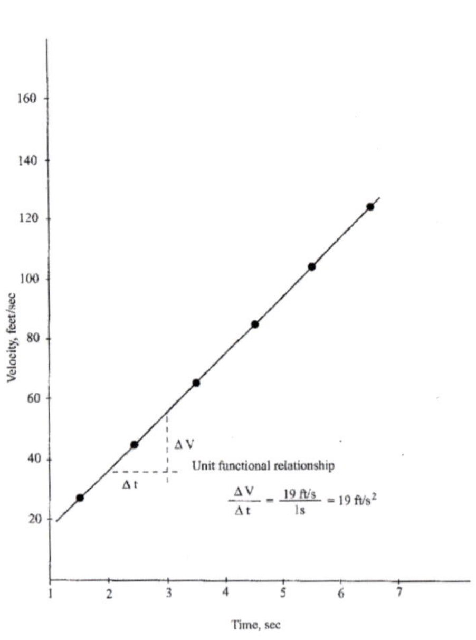

Figure 4.27: Learner's graph of time-velocity relationship the object has fallen

Chapter Five

Applying Cognitive Functions to Achieving Mathematical Academic Standards

What are Mathematics Standards?

Mathematics equips people to think about the patterns and relationships in the world of objects and events and to communicate these thoughts with order, coherency, and logic. Mathematical thinking, like language, is an intrinsic aspect of our cultural system, and it helps to drive and shape scientific discoveries, medical advancements, and technological innovations. Dr. Michael Brown, Nobel Prize Laureate, stated: "If America is to maintain our high standard of living, we must continue to innovate. We are competing with nations many times our size. We don't have a single brain to waste. Math and science are the engines of innovation. With these engines we can lead the world. We must demystify math and science so that all students feel the joy that follows understanding." (Colorado Department of Education, n.d.)

Mathematics comprises the processes, tools, and language that help us to engage in formulating, organizing, interpreting, hypothesizing about, and validating quantitative information. Thus, a mathematically informed citizenry is essential for a thriving democracy and a technologically driven economy. This establishes the rationale and need for robust mathematics education standards in the United States.

Mathematical standards specify what students should understand and be able to do in their study of mathematics. In the RMT paradigm, learners' understanding is demonstrated through their providing mathematical logical evidence to validate their findings, justify their lines of inquiry, and create new insights.

Below are two National Council of Teachers of Mathematics (NCTM) algebra standards with their grade level expectations (National NCTM, 2000)

First Example of NCTM Algebra Standard

Instructional programs from prekindergarten through grade 12 should enable each student to understand patterns, relations, and functions

Grades 3–5 Expectations: In grades 3–5 each student should–

- describe, extend, and generalize about geometric and numeric patterns;

- represent and analyze patterns and functions, using words, tables, and graphs.

Grades 6–8 Expectations: In grades 6–8 each student should–

- represent, analyze, and generalize a variety of patterns with tables, graphs, words, and, when possible, symbolic rules;
- relate and compare different forms of representation for a relationship;
- identify functions as linear or nonlinear and contrast their properties from tables, graphs, or equations.

Second Example of NCTM Algebra Standard
Analyze change in various contexts

Grades 3–5 Expectations: In grades 3–5 each student should–

- investigate how a change in one variable relates to a change in a second variable;
- identify and describe situations with constant or varying rates of change and compare them.

Grades 6–8 Expectations: In grades 6–8 each student should–

- use graphs to analyze the nature of changes in quantities in linear relationships.

Grades 9–12 Expectations: In grades 9–12 each student should–

- approximate and interpret rates of

change from graphical and numerical data.

Below are two examples of the Common Core State Standards on the big idea of ratios, rates, and proportions.

CCSS.Math.Content.6.RP.A.3

Use ratio and rate reasoning to solve real-world and mathematical problems, e.g., by reasoning about tables of equivalent ratios, tape diagrams, double number line diagrams, or equations.

CCSS.Math.Content.7.RP.A.3

Analyze proportional relationships and use them to solve real-world and mathematical problems.

In the next section, we present example scenarios on learners' application of cognitive functions that achieve mathematical academic standards through problem solving. Some of these problems are modifications of PARCC Released Items.

Applying Cognitive Functions to Solving Standards-based Mathematics Problems

Problem #1

Ryan and Jay started walking at the same time using the same path. Ryan walks from Pine Grove to Dutch Fork with a constant speed of 3.0 miles per hour, and Jay from Dutch Fork to Pine Grove with a constant speed of 2.5 miles per hour. They meet in two hours. The path each takes runs northward from Pine Grove to Dutch Fork. At what point along the path do they meet? Is their distance of separation proportional to the remaining time for walking before meeting? What is the distance from Pine Grove to Dutch Fork? Provide mathematical logical evidence to validate your answer.

<u>Problem's Cognitive Demand</u>

Analyzing, visualizing, labeling, comparing, forming relationships, and integrating are required to understand what is presented in the verbal presentation of the problem.

Defining the problem is needed to determine and specify the known and the unknown, and to construct one or more logical pathways to move from the known to the unknown.

Considering more than one source of information simultaneously, selecting relevant cues, and providing mathematical logical evidence are required to support defining the problem.

In RMT, learners are mediated to construct their own diagrams to assist them in unpacking the structure of the problem and visualizing the flow of its components. This helps learners in building their understanding of what is given and perceiving relational aspects

Using a Diagram to Unpack the Structure of the Problem

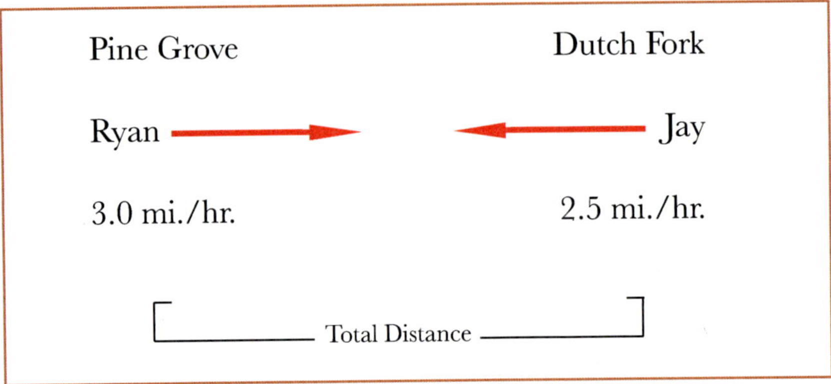

of the mathematical event given within the verbal presentation of the problem. Planning and drawing a diagram engage learners in performing a structural and operational analysis of the problem. Figure 5.1 presents a learner's approach to analyzing, defining and solving the problem.

Below, we present the results of a class-room of learners' collective work to define and solve the problem.

Defining the problem

Encoding the speed of Ryan	V_r
Encoding the speed of Jay	V_j
Encoding time	t
Encoding distance	d

Known	Unknown
$V_r = 3.0$ mi/h	dtotal = ?
$V_j = 2.5$ mi/h	Meeting Point = ?
$t_{final} = 2.0$ h	Is d_s proportional to t?

Constructing a Logical Pathway from Known to Unknown

Selecting as a relevant cue "They meet in two hours" and activating prior knowledge that $d = v \cdot t$ positions us to form the quantitative relationships:

$$d_r = V_r \cdot t \text{ and } d_j = V_j \cdot t$$
$$d_{total} = V_r \cdot t + V_j \cdot t = 3.0 \text{ mi/h} \times 2 \text{ h}$$
$$+ 2.5 \text{ mi/h} \times 2 \text{ h} =$$
$$6.0 \text{ mi} + 5.0 \text{ mi} = 11.0 \text{ mi}$$

Ryan and Jay will meet on the path at a point that is 6.0 miles north of Pine Grove and 5.0 miles south of Dutch Fork.

Is the distance of separation between Ryan and Jay proportional to the remaining time for walking before meeting? If d_s is proportional to trem, d_s/trem will conserve constancy for all corresponding quantities, amounts, or magnitudes of the variables trem and d_s. Let us use the structure of the ta-

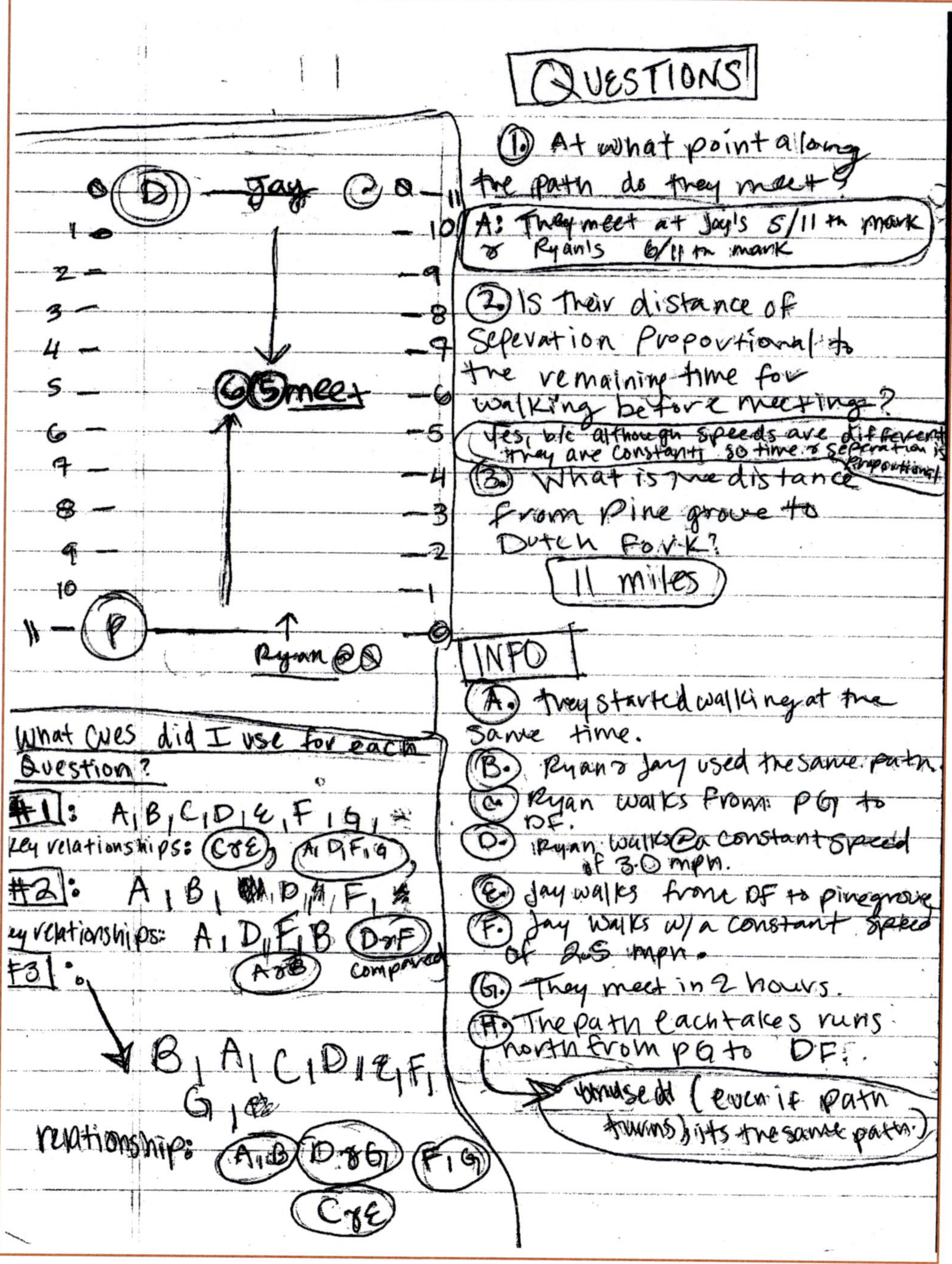

Figure 5.1: A learner's approach to analyzing and defining the problem

Table 5.1—Forming Proportional Quantitative Relationships between Distance Separated and Remaining Walking Time

Time, t (hour)	Time Remaining, t_{rem} (hour)	Distance Walked, d_w (mile)	Distance Separated, d_s (mile)	Ratio d_s/t_{rem} (mile/hour)
0.25	1.75	1.37	9.63	5.5
0.5	1.5	2.75	8.25	5.5
1.0	1.0	5.5	5.5	5.5
1.5	0.5	8.25	2.75	5.5
1.75	0.25	9.62	1.37	5.5

ble on page 162 as a mathematical psychological tool to construct a system of quantitative relationships, as ratios, and search for mathematical logical evidence of proportionality.

The following are supporting quantitative relationships:

$$t_{rem} = 2.0\,h - t; \quad d_w = v_R \cdot t + v_J \cdot t; \text{ and,}$$
$$d_s = 11.0 - d_w.$$

Since the ratio of distance separated and time remaining for walking conserves constancy with a value of 5.5 miles/hour for all times of walking, d_s is proportional to t_{rem}. The two variables are forming proportional quantitative relationships.

We can shift the perspective of how these two variables are interacting through the cognitive function forming a functional relationship between two variables. Notice that this is an example where problem solving in RMT aims toward a broader and more transcendent scope than simply producing an answer. Exploring how big ideas intersect within the specifics and context of a single problem provides deeper insight into the systemic nature of mathematics concepts. Here learners experience the dynamic conceptual interrelatedness of proportional quantitative relationships and a functional relationship between variables.

When forming a functional relationship between two variables, learners are creating an understanding of how changes in the quantity or amount or value or magnitude of an independent variable or cause variable or input variable bring about corresponding changes in the quantity or amount or value or magnitude of a dependent variable or an effect variable or an output variable. We can utilize the mathematical psychological tools of a table and the x-y coordinate plane to construct this functional relationship between the variables t_{rem} and d_s. The distance of separation between Ryan and Jay appears to depend directly on the remaining time they have to walk before meeting.

Table 5.2—Forming a Functional Relationship between Distance Separated and Remaining Walking Time

Independent Variable, t_{rem} (hour)	Dependent Variable, d_s (mile)	Ordered Pair	Change in d_s, Δd_s (mile)	Change in t_{rem}, Δt_{rem} (hour)	Rate of Change $\Delta d_s/\Delta t_{rem}$ (mile/hour)
1.75	9.62	(1.75, 9.62)			
			-1.37	-0.25	5.5
1.50	8.25	(1.50, 8.25			
			-2.75	-0.50	5.5
1.0	5.5	(1.00, 5.50)			
			-2.75	-0.50	5.5
0.5	2.75	(0.5, 2.75)			
			1.38	-0.25	5.5
0.25	1.37	(0.25,1.37)			

In this table, the functional relationship between the two variables is conveyed in the third column containing the ordered pairs. As we move vertically down the column, the quantity of the dependent variable, the distance of separation between Ryan and Jay, decreases as the quantity of the remaining time of walking decreases.

In the RMT paradigm, the learner is mediated to build the cognitive function, forming a unit functional relationship between the two variables. By executing this thinking action, the learner understands the quantity or amount or value or magnitude of change in the dependent variable or the effect variable or the output variable that is brought about by a unit change in the quantity or amount or value or magnitude of the independent variable or the cause variable or the input variable. The rate at which this occurs is shown in the last column, and is the combined speed of the two walkers.

Problem #2

The first term in a sequence is b. If every term thereafter is 5 greater than 1/10 of the preceding term, and b ≠ 0, what is the ratio of the second term to the first term when it is written as an algebraic expression? Provide all your mathematical logical evidence.

a) (b + 50)/10b
b) (b + 10)/5
c) (b + 5)/10
d) (b + 10)/50b
e) None of the above

Defining the Problem

Known	Unknown
1st term = b	Ratio of 2nd term/1st term = ?
2nd term = 1/10 b + 5	

The logical pathway from the known to the unknown is forming a quantitative relationship, whose structure is presented in the unknown. Ratio of 2nd term/1st term = ((1/10) b + 5)/b = (1/10) + 5/b = (b + 50)/10b, which is choice a) above. Let us begin providing mathematical logical evidence to justify this result by considering the property that a quantity, multipled by its reciprocal, is equal to one. Given that d/c is the reciprocal of c/d.

(c/d) · (d/c) = (c · d)/(d · c)

We know from the commutative law of multiplication that c · d = d · c; thus (c · d)/(d · c) = 1. If we take the ratio of the first term in the sequence to the second term, this will be the reciprocal of the second term to the first term.

By forming a new quantitative relationship, we produce the ratio of 1st term to 2nd term = b/((1/10)b + 5) = b/((b + 50)/10) = 10b/(b + 50). If this is indeed the reciprocal of result we produced as the solution to the original problem, which was the ratio of the second term to the first term, multiplying one times the other will produce a product equal to one.

(10b/(b + 50)) · (b + 50)/10b = ((10b · (b + 50))/((b + 50) · (10b)) = 1. Thus, we have dynamically used cognitive functions to verify our solution to the original problem.

A learner independently applies cognitive functions to solve this problem (see Figure 5.2).

Problem #3

Cheetahs are among the fastest land animals, with a top speed of around 88 feet per second. However, because of their physiology and the challenges of their habitat, maintaining this top speed beyond 20 seconds causes over-exertion and over-heating to their brain, lungs, and other organs. On the other hand, a choice prey of the cheetah, the wildebeest, has a top speed of 51 feet per second with considerable endurance and stamina. What distance does the wildebeest need to maintain from the cheetah to be safe?

An approach to analyzing this problem and designing a solution strategy is given by Craig Washington, an RMT Laboratory learner. We begin with the cognitive function defining the problem. We must determine and identify, clearly and precisely, what is known and what is unknown, and construct one or more logical pathways to connect the known with the unknown. The cognitive function analyzing is clearly needed to begin this process.

Analyzing the verbal presentation of the problem:

1. Cheetah can run 88 ft/s for 20 s.
2. Wildebeest can run 51 ft/s for a much longer period of time.

Encoding the speed of the cheetah v_C
Encoding the speed of the wildebeest v_W

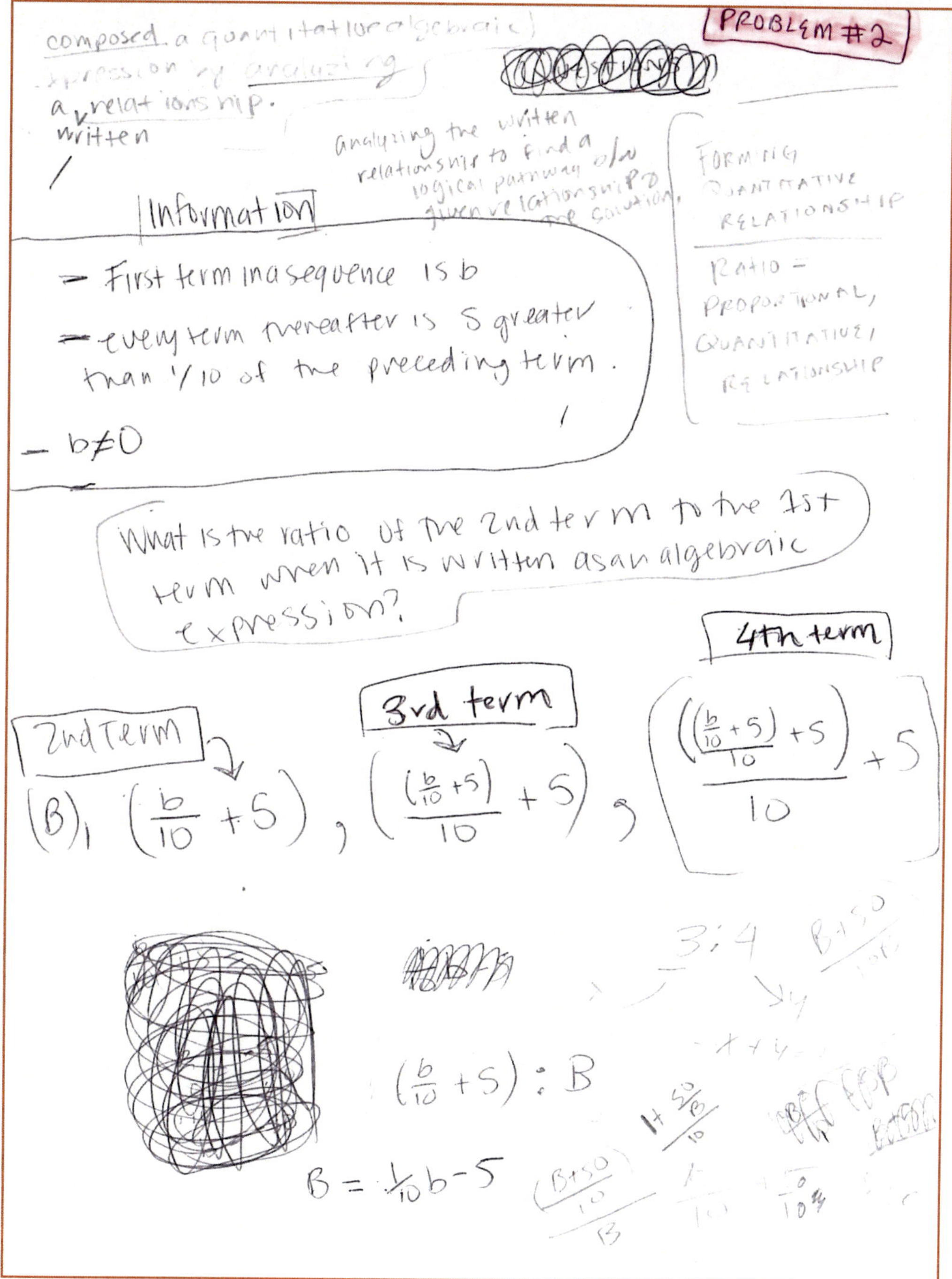

Figure 5.2: A learner's independent work on applying cognitive functions to solve the problem

Drawing a Diagram to Analyze the Structure of the Problem

At t = 0 second

0

j ....

Initial quantity of linear space that is safe for the wildebeest to be from the cheetah.

Encoding the time of the chase — t

Encoding the quantity of linear space or distance of chase where wildebeest can be caught — j

Encoding the safe distance the wildebeest must maintain from the cheetah before the chase — d_s

Known	Unknown
$v_c = 88$ ft/s for 20 s	$d_s = ?$
$v_w = 51$ ft/s	

Constructing a Logical Pathway from Known to Unknown

Comparing the speed of the cheetah with the speed of the wildebeest, we recognize that the cheetah is much faster than the wildebeest for the first 20 seconds of the chase. Selecting the relevant cue of 20 seconds of chase time for the cheetah to be able to catch the wildebeest and activating the prior knowledge that speed x time = distance, we form the quantitative relationships for the first 20 seconds of the chase:

$$d_s \geq j; \quad d_w = v_w \cdot 20 \text{ s} + j; \quad d_c = v_c \cdot 20 \text{ s}; \quad \text{and} \quad d_w = d_c$$

Thus,

$$v_w \cdot 20 \text{ s} + j = v_c \cdot 20 \text{ s}$$
$$51 \text{ ft/s} \cdot 20 \text{ s} + j = 88 \text{ ft/s} \cdot 20 \text{ s}$$
$$1020 \text{ ft} + j = 1760 \text{ ft}$$
$$j = 1760 \text{ ft} - 1020 \text{ ft} = 740 \text{ ft}.$$

$d_s \geq 740$ ft. Thus, the wildebeest needs to maintain a quantity of linear space or distance equal to or greater than 740 feet from the cheetah to be safe.

Providing Mathematical Logical Evidence to Support this Result

We can provide mathematical logical evidence to support or validate our result above by viewing d_w and d_c as variables, each of which, is forming a functional relationship with t, the time of the chase from 0–20 seconds. The generalized algebraic equations that express these two functional relationships are:

1) $d_W = v_W \cdot t + j = 51$ ft/s $\cdot$ t $+ 740$ ft
2) $d_C = v_C \cdot t = 88$ ft/s $\cdot$ t

We can use a table and the x-y coordinate plane as mathematical psychological tools to solve this system of equations.

The functional relationship between the independent variable, t, and the corresponding dependent variable, d_W, is presented in the third column, while the functional relationship between t and the other corresponding dependent variable, d_C, is conveyed in the fifth column. In the third column, we observe that as the quantity or amount or value or magnitude of time increases, the corresponding quantity or amount or value or magnitude of the distance the wildebeest runs also increases. A similar pattern appears in the fifth column, showing that as the time of the chase increases the corresponding distance the cheetah runs also increases.

Notice that at t_0, the time just at the start of the chase, the wildebeest is 740 feet from the cheetah, while the cheetah's starting position is 0 feet, establishing the spatial frame of reference for the linear motion of the two animals. Just after a 20-second chase with both animals running at their top speed, the cheetah will approach catching the wildebeest at the cheetah's distance of running equal to 1760 feet, which is the same distance the wildebeest will be from the initial starting point of the cheetah. But at this point in time and distance of running at top speed, the cheetah will be exhausted, and the wildebeest will continue into safety. This provides the mathematical logical evidence $d_S \geq 740$ ft, which means that the wildebeest needs to maintain a quan-

Table 5.3—Forming a Functional Relationship between Time of Chase and the Distance the Wildebeest Runs and Time of Chase and the Distance the Cheetah Runs

Time, t (Seconds)	Distance of Wildebeest d_W (Feet)	Ordered Pair (t, d_w)	Distance of Cheetah d_w (Feet)	Ordered Pair (t, d_c)
Independent Variable	Dependent Variable		Dependent Variable	
0	740	(0, 740)	0	(0, 0)
5	995	(5, 995)	440	(5, 440)
10	1,250	(10, 1250)	880	(10, 880)
15	1,505	(15, 1505)	1320	(15, 1320)
20	1,760	(20, 1760)	1760	(20, 1760)

tity of linear space or distance equal to or greater than 740 feet from the cheetah to be safe.

We can provide this mathematical logical evidence in a more explicitly visual form by forming two functional relationships between the two variables t and d_w and between the two variables t and d_c. The graphs in Figure 5.3 show that although the wildebeest starts running at 740 feet ahead of the cheetah, the greater rate of change in distance per second for the cheetah allows the cheetah to approach catching the wildebeest at 20 seconds and after running 1,760 feet. This greater rate of change in distance per unit time of the cheetah versus the wildebeest is visually captured by comparing the slopes of the graphs.

As stated earlier, in the RMT paradigm the learner is mediated to build the cognitive function, forming a unit functional relationship between two variables. By executing this thinking action, the learner understands the quantity or amount or value or magnitude of change in the dependent variable or the effect variable or the output variable that is brought about by a unit change in the quantity or amount or value or magnitude of the independent variable or the cause variable or the input variable. The slope or steepness of the graph expresses the quantity of change in distance per quantity of change in time, which is a ratio. For each graph, this ratio conserves constancy in its value for a progressive series of ratios, and is thus forming a system of proportional quantitative relationships.

It is evident that through the RMT paradigm, learners build and dynamically apply cognitive functions to construct mathematical conceptual understanding that creates new insight into systemic relationships between basic math (ratios and proportions) and algebra (rates, slopes, and functional change).

From the table, construction of a graph plotting Distance (feet) vs. Time (seconds) for each animal can be accomplished.

From observation, the plots of the two animals are linear, as would be expected because the equations from which the functional relationships are derived for the two animals are in the form of the equation of a line: $y = mx + b$ …

where "y" is the dependent variable, "m" is the slope of the line, "x" is the independent variable and "b" is the y-intercept.

Distance of Cheetah: $d_c = v_c \cdot (t) = 88$ ft/sec x elapsed sec

Distance of Wildebeest: $d_w = v_w \cdot (t) + j = 51$ ft/sec x elapsed sec + 740 ft "headstart"

When merely observing, one will notice the steepness of the line representing the

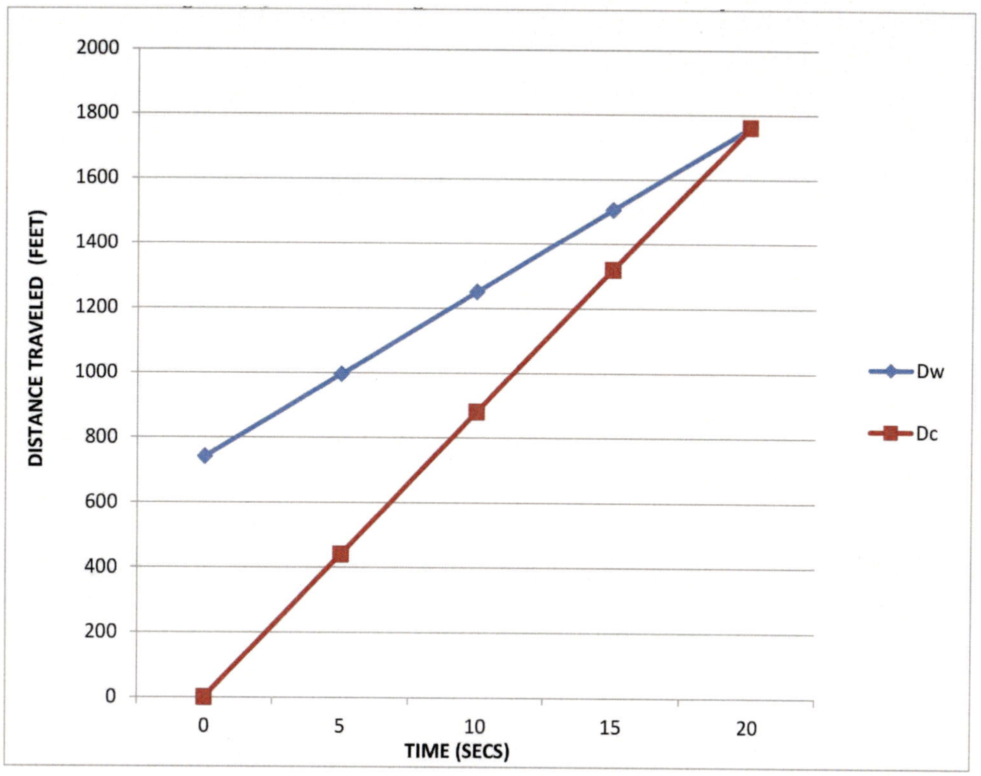

Figure 5.3—Time of Flight vs. Distance Traveled by Wildebeest Compared to Time of Chase vs. Distance Traveled by Cheetah

cheetah as compared to the wildebeest. From prior knowledge, we know this is due to the slope of the line, the adage "rise over the run," which in this case is the "speed" or "velocity" of the cheetah at 88 ft/sec.

Slope "m" $= (\Delta y/\Delta x)$
which is (change in y/change in x)

$\underline{\text{change in distance (feet)}}$.
change in "time" (sec)

Velocity is the directional distance displaced per unit of time. Using two data points from the table, the slope of the lines can be calculated and confirmed.

For the cheetah,
the slope $= \dfrac{440 - 0 \text{ (feet)}}{5 - 0 \text{ (sec)}} = \dfrac{440 \text{ (ft)}}{5 \text{ (sec)}} =$

$88 \text{ ft/sec} = v_C$

For the wildebeest,
the slope $= \dfrac{995 - 740 \text{ (feet)}}{5 - 0 \text{ (sec)}} = \dfrac{255 \text{(ft)}}{5 \text{ (sec)}}$

$= 51 \text{ ft/sec} = v_W$

The slope of the line for the cheetah tells us that its distance rate of change (w/

respect to time) or "speed" or "velocity" at 88 ft/sec is much faster than the wildebeest at 51 ft/sec.

Further observance of data in the table reveals that the slope of each line for each animal remains constant or conserves constancy over the 20-second chase.

In conclusion, applying the cognitive functions forming a unit functional relationship, defining the problem, providing mathematical logical evidence, selecting relevant cues, activating prior knowledge, and using mathematically specific psychological tools enabled a way to visually and graphically analyze and characterize the chase of the cheetah for a wildebeest dinner. Although the graph clearly shows the cheetah is the faster animal and did catch up to the slower wildebeest, it did not make the kill for 2 reasons:

1. The cheetah can only give 20 seconds of chase at top speed!
2. The wildebeest maintained a 740 ft. headstart long enough to poop the cheetah out! Sorry cheetah, no wildebeest chops, steaks, hams and tenderloins for you today!

In Figure 5.4, another learner constructs a diagram that serves as her idosyncratic psychological tool to help her perform both a structural analysis and operational analysis of this problem. She displays her strategic competence in the underlying big mathemaical ideas, as well as her adaptive reasoning in her presentation of "missing evidence" and "assumptions."

Problem #4

A ball always bounces back to 2/5 of the height of its previous bounce after being dropped. After the first bounce, it reaches a height of 175 inches. Approximately, how high (in inches) will it reach after the third bounce? Provide all your mathematical logical evidence to justify your response.

In Figure 5.5, a learner unpacks the structure of the problem by performing a structural and operational analysis of it. Notice that her deep conceptual understanding and productive disposition compel her to go beyond just getting the answer to the problem. She derives a general formula to predict the height the ball bounces and elaborates on it limitations.

Problem #5

The flow of water through a certain pipe is 20 cubic meters per minute. How many minutes would it take for four such pipes to fill 2 tanks, if each tank is a cube with a side length of 20 m? Provide all your mathematical logical evidence to support each step in your strategy and to justify your results.

Defining the problem

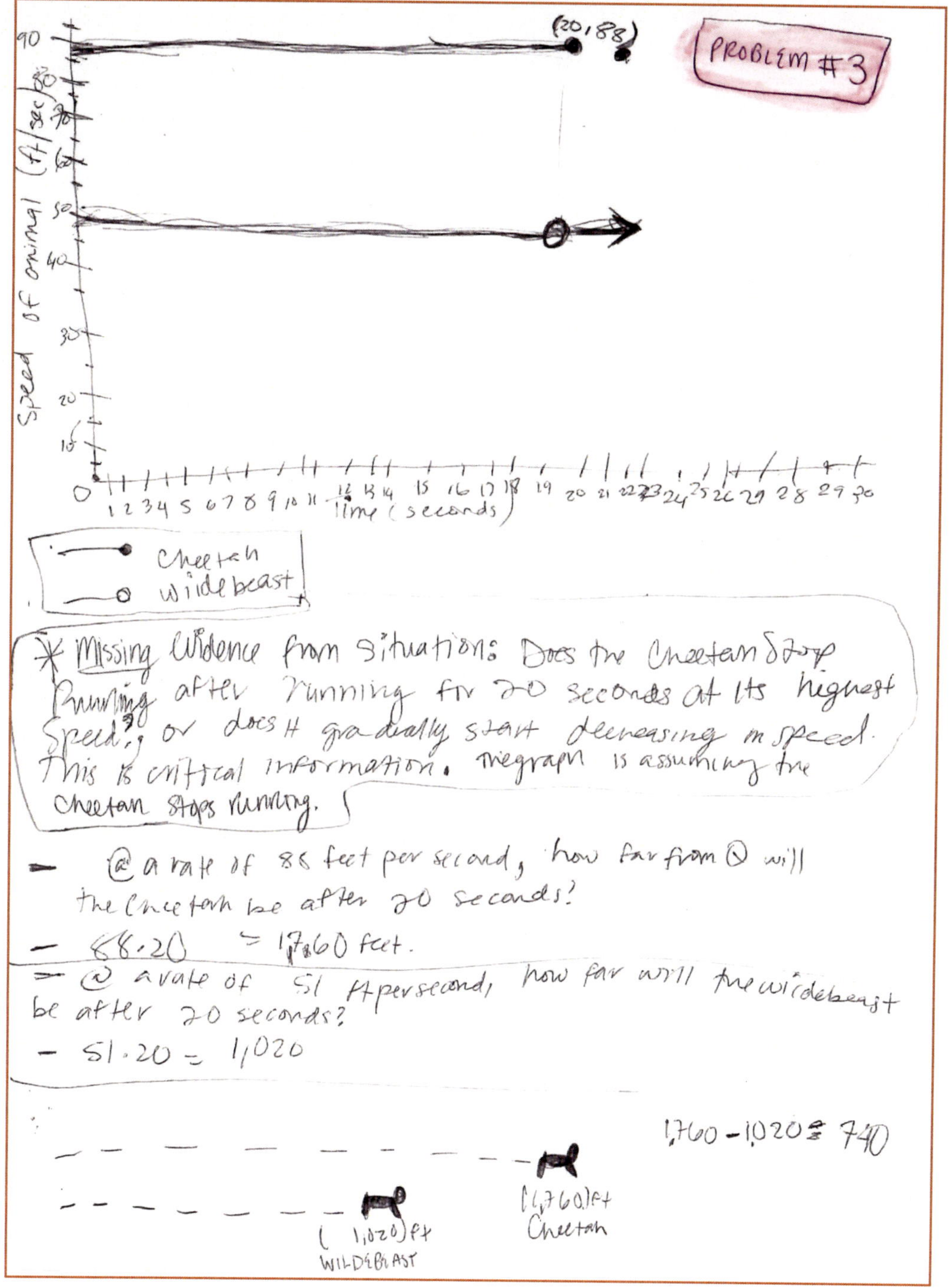

Figure 5.4: A learner's independent work on problem #3

assuming the cheetah stops running after 20 seconds and assuming the animals started running at their maximum speed, The Cheetah will be positioned at 1760 feet after 20 seconds & the wildebeast at 1020 feet.

In order to stay safe from the cheetah, the wildebeast must stay a distance longer than 740 feet away from the cheetah.

• analyzing   • Integrating   • Forming relationships • Focusing

INFO
(A) Cheetah's top speed is around 88 feet per second.
(B) Cheetah's can maintain top speed for 20 seconds
(C) Wildebeast have a top speed of 51 ft per second
(D) Wildebeast have considerable endurance & stamina.

QUESTION: What distance does the wildebeast need to maintain from the cheetah to be safe?

↳ Info used – A, B, C, D.
relationships formed – (A&B), (C&D), (A&B) & (C&D).

• I analyzed the information from the problem and integrated it in the form of a graph. The graph shows the speed maintained by the cheetah & wildebeast over 20 seconds, assuming that both animals started from zero at their highest speed. b/c cheetah's can only maintain their speed for 20 seconds, I assumed that the cheetah stopped moving at 20 seconds. I assumed that the wildebeast could maintain the highest speed after 20 seconds.

I further analyzed the information from the graph which comprised of the given info. from the situation and the information assumed based on the given information. I integrated this into another visual that showed the distance from zero (in feet) each animal would be after 20 seconds of full speed running. I was able to find these values by identifying the relationship between time and speed of each animal, in the situation), at a constant speed, the distance from zero can be found by multiplying the rate per second of movement by the # of seconds of movement.

The value 740 was calculated by subtracting the wildebeasts distance at 20 seconds from the cheetah's distance from zero at 20 seconds. I formed the relationship between the two distance values of each animal. If the cheetah is not to catch the wildebeast, then at 20 seconds. The value of the wildebeast's distance must be greater than that of the cheetah. If both animals started at the distance zero, then after 20 seconds the wildebeast is 740 feet short of the cheetah's distance. This means that the cheetah will have caught the wildebeast nearly immediately. Instead, the cheetah if starting at zero, will only miss the wildebeast if the wildebeast starts at a distance greater than 740 ft from zero.

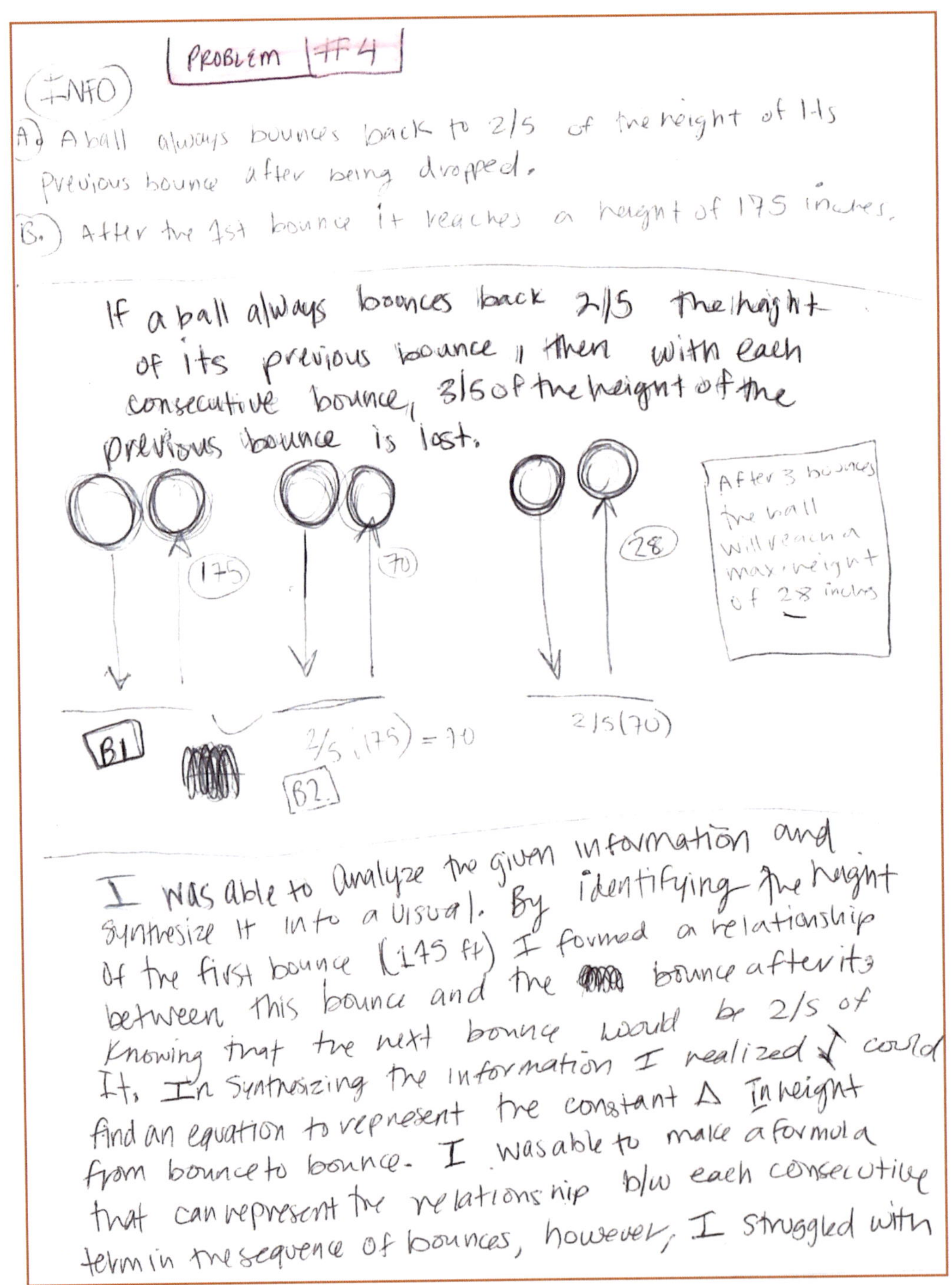

Figure 5.5: A learner's independent work on applying cognitive functions to understanding and solving problem 4

Writing an equation that could ~~~~ easily find the height of bouncing by plugging in the # term. The expression I found is: $\left(\frac{2}{5}\right)y_1 = y_2$, $\left(\frac{2}{5}\right)y_2 = y_3$ and so on. This works as a formula to find bounce height only when the value of the previous bounce height is known, however you can not use this formula to to find term y_n without knowing the value of y_{n-1}. The expression can be written as $\left(\frac{2}{5}\right)y_{(n-1)} = y_n$

The next step in finding an equation to this problem is finding a way to find the value of $y_{(n-1)}$ for each value of n. In the equation, I do know that "Height Dropped from" is a variable, because all other values of y ~~are based~~ depend on the original height from which the ball is dropped.

Known

$v = 20 \text{ m}^3/\text{min/pipe}$
$V = (20\text{m})^3/\text{tank}$
tanks = 2

Unknown

t (time) to fill 2 tanks =?

In Figure 5.6, a learner applies cognitive functions to follow a logical pathway that connects the known with unknown. Notice that the learner extensively uses the cognitive function forming relationships to provide mathematical logical evidence and express conceptual understanding.

Problem #6

Use the information provided to answer Part A and Part B for problem 6.

Tevera has to drive her car as part of her job. She receives money from her company to pay for the gas she uses. The table shows a proportional relationship between y, the amount of money that the worker receives, and x, the number of work-related miles driven.

Part A

Explain how to compute the amount of money Tevera receives for any number of work-related miles. Based on your explanation, derive an equation that can be used to determine the total amount of money, y, Tevera receives for driving x work-related miles.

Distance Driven, x (miles)	Amount of Money Received, y (dollars)
25	12.75
35	17.85
40	20.40
50	25.50

Defining the problem

Known
x = amount of work-related driving miles
y = amount of money Tevera receives for a specific amount of work-related miles

Unknown
Relationship between y and x =?

Constructing Logical Pathway that Connects the Known to the Unknown

The pathway begins with expanding the given table in the problem and using it as a mathematical psychological tool to execute the three cognitive functions:

a) Forming proportional quantitative relationships;
b) Forming a functional relationship between two variables; and
c) Forming a unit functional relationship.

When forming proportional quantitative relationships, one is using his or her mind to construct a system of proportionality (proportions) in which corresponding quantities (amounts or magnitudes) of two variables form a progression of ratios that are all equivalent or conserve constancy in their intrinsic value. The third column in the expanded table below shows that the ratio of the amount of money Tevera receives to the number of work-related mi conserves constancy with a value of $ 0.51/mile. In order to compute the amount of money Tevera receives, y, for any number of work-related miles, x, we form the ratio and proportion relationship below:

$$(y/x) = \$\, 0.51/\text{mile.}$$
$$y = (\$\, 0.51/\text{mile}) \cdot x$$

The two variables are also interacting by forming a functional relationship with each other, with the number of work-related miles, x, being the independent variable and the amount of money Tevera receives, y, being the dependent variable. In the last column, we are forming a unit functional relationship by determining that the quantity of change in the money Tevera receives that is produced by a unit change in work-related miles is $0.51/mile. This is the rate of change in y per change in x, which is the same as the ratio of y/x.

Again, we have shown that through the RMT paradgm, learners can dy-

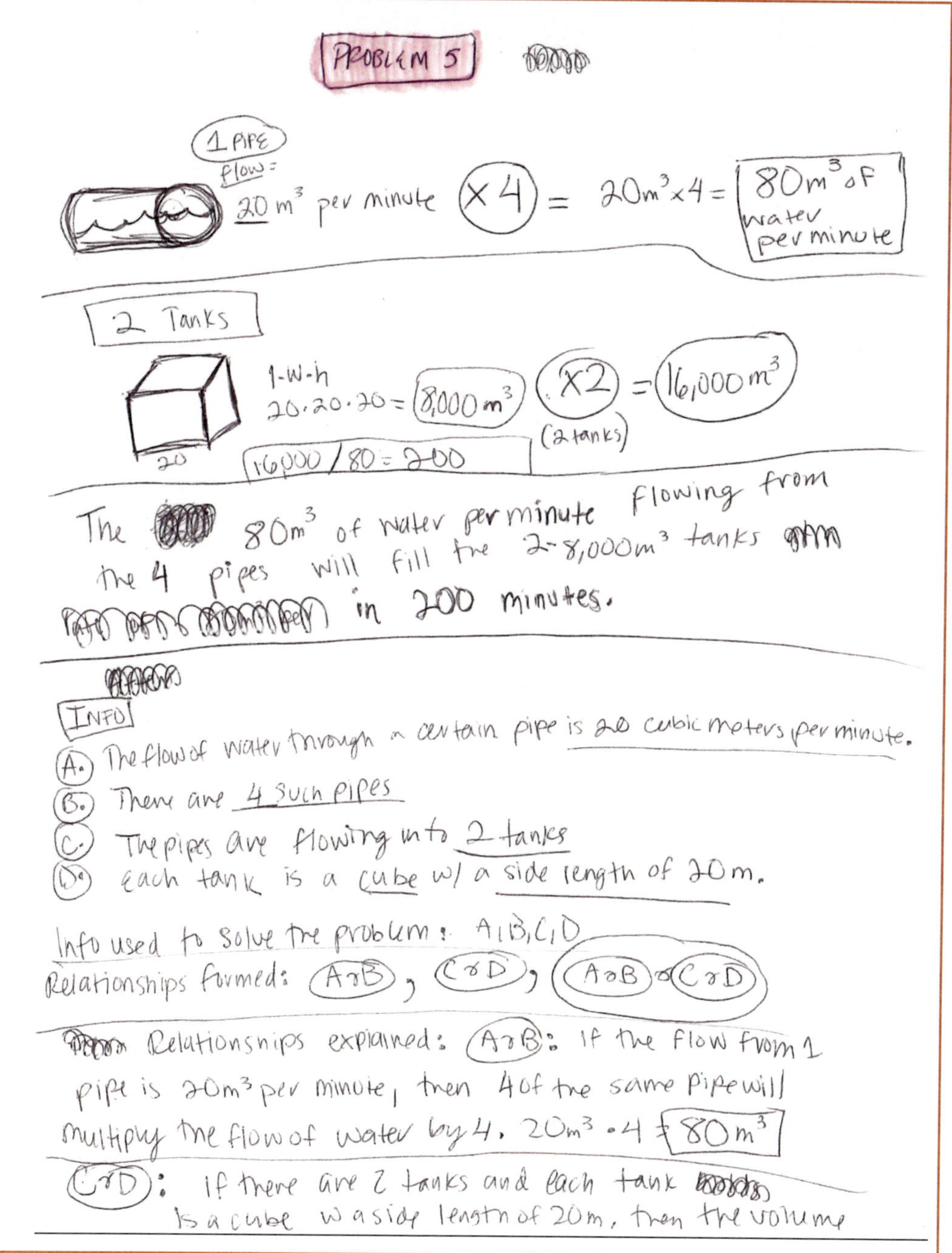

Figure 5.6—A learner's independent work in applying cognitive functions to understand and solve problem #5

namically apply cognitive functions to constructing mathematical conceptual understanding that creates new insight into systemic relationships between basic math (ratios and proportions) and algebra (rates, slopes, and functional change). The cluster of cognitive functions that are building this structure of thinking consists of comparing, integrating, conserving constancy, forming proportional quantitative relationships, forming a functional relationship between two variables, and forming a unit functional relationship.

Part B

On Monday, the worker drove a total of 134 work-related and personal miles. She received $32.13 for the work-related miles she drove on Monday. What percent of her total miles driven were work-related on Monday? Show or explain your work.

Known

$y = \$32.13$

$x + personal = 134$ miles

Unknown

% $(x)/(x + personal) \cdot 100 = ?$

Constructing Logical Pathway from Known to Unknown

$y = (\$\,0.51/miles) \cdot x$

$x = y/(\$0.51/miles) = \$32.13/(\$0.51/miles) = 63$ miles

% $(x)/(x + personal) \cdot 100 = (63$ hours$/134$ hours$) \cdot 100 = 47$ %.

Problem #7

Lateshia is driving on a long straight road that inclines at an angle of 8.6^0 above the horizontal. She observes a sign at one point that states "Elevation 385 m." What is her elevation after she has driven another 3.0 km? What is the horizontal distance covered by her car? Learners begin by reading and analyzing

Table 5.4—Expanded Table

Work-related miles, x Independent Variable,	Money Tevera Receives y ($) Dependent Variable,	Ratio of y/x ($/mile)	Ordered Pair	Change in y, Δy ($)	Change in x, Δx (mile)	Rate of Change Δy/Δx ($/mile)
25	12.75	0.51	(25, 12.75)	5.10	10	0.51
35	17.85	0.51	(35, 17.85)	2.55	5	0.51
40	20.40	0.51	(40, 20.40)	5.10	10	0.51
50	25.50	0.51	(50, 25.50)			

the problem. They capture and show the relationships among the components of their analysis with a diagram. One example is presented below.

Using a Diagram to Unpack the Structure of the Problem

According to the RMT paradigm, every quantity has a defining property, which lends itself to precise measurement by an appropriate tool. This defining property is theoretical in nature and is independent of specific objects. A conceptual analysis of the problem is given below.

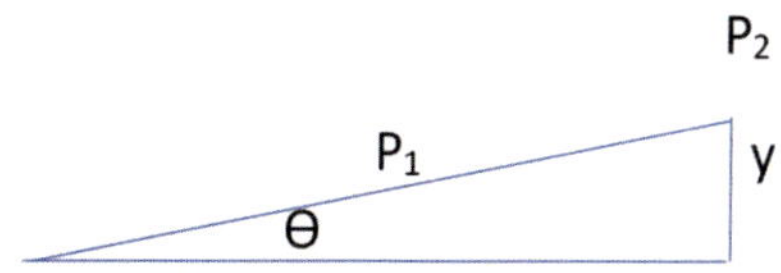

Conceptual Analysis of the Problem

A. Quantities of Linear Space
1. A long straight road
2. Elevation of 385 m at Point P_1
3. 3.0 km from P_1 to P_2
4. Elevation at P_2
5. Horizontal distance covered by car

B. Quantity of Rotation of a Line Segment (measure of angle of incline)
1. Road inclines at an angle of 8.6^0

There are five quantities of linear space identified in this analysis. The defining property for linear space is "a straight path between two endpoints." The quantity of space in a linear path is between two points is what is sought in "What is the horizontal distance covered by the car?" and what is given in "elevation of 385 m," etc.

Defining the problem

Known
Elevation at P_1 = 385 m
Distance from P_1 to P_2 = 3.0 km
Angle of incline: m θ = 8.6^0

Unknown
Elevation at P_2 =?
Horizontal distance covered =?

Constructing a Logical Pathway from the Known to the Unknown

1. Applying geometric concepts, let us construct a line segment that begins at P_2 and connects with and forms a perpendicular to the horizontal path. This creates a right triangle that is bounded by the total distance on the incline, c, the total distance of the horizontal path, x, and the final elevation, y.

2. Next, we turn to the basic trigonometric function sin θ = (quantity of linear space in opposite side)/(quantity of linear space in the hypotenuse) = y/c.

3. Rearranging: $y = c \sin \theta = c \sin 8.6 = 0.1495\ c.$

4. Let us turn back to geometry and construct a line segment that begins at P_1 and connects with and forms a perpendicular to the horizontal path at x_1. This creates a right triangle, bounded by the distance from the vertex of angle θ to P_1 on the incline, the elevation from x_1 to P_1, which is the elevation of 385m, and the distance from the vertex of angle $\angle\theta$ to x_1 on the horizontal path.

5. Turning again to the basic trigonometric function $\sin \theta = $ (quantity of linear space in opposite side)/(quantity of linear space in the hypotenuse) $= 385m/c_1$: $c_1 = 385m/(\sin \theta) = 385/0.1495 = 2.6$ km.

6. We can now determine c, the total quantity of linear space along the incline from the vertex of angle$\angle\theta$ to point P_2. $c = 2.6$ km $+ 3.0$ km $= 5.6$ km or 5,600 m.

7. Now we can determine y, the elevation at P_2, by substituting the value for c in the equation we derived in 3 above: $y = c \sin \theta = c \sin 8.6 = 0.1495$ c. Thus, $y = 0.1495$ x 5,600 m $= 837$ m.

8. When we focus on the larger right triangle constructed in 1 above, along with the basic trigonometric function $\tan \theta = $ (quantity of linear space in the opposite side)/(quantity of linear space in the adjacent side),

we have the equation:

$\tan \theta = y/x$. Rearranging to determine x, the total horizontal distanced traveled, we have: $x = y/\tan \theta = 837$ m/$0.1512 = 5,536$ m.

Reflections and Insights

The basic trigonometric function $\sin \theta$ is a ratio between two quantities of linear space—the (quantity of linear space in the opposite side)/(quantity of linear space in the hypotenuse). At any point along the incline, we can construct a perpendicular to the horizontal path and create a new right isosceles triangle. The ratios of all such triangles formed this way belong to, and form, a system of proportional quantitative relationships, where the proportionality constant is, in this case, 0.1495.

If we consider the movement from the vertex of the angle to point P_2 as a mathematical event, x and y become the algebraic concepts of variables that form a functional relationship with each other. The rate of change in the dependent variable, y, that is produced by a unit change in the independent variable, x, is the slope of the incline, which is the tangent of angle of incline, θ. This rate of change is forming a unit functional relationship, which is $\Delta y/\Delta x$.

Suppose the angle of incline on the road Lateshia is driving on starts changing at some point P_3. At this point, we must shift our theoretical thinking and our mathematical apparatus, for $\Delta y/\Delta x$ begins changing itself. Our point of reference must shift to express the nature of how the dependent variable, y, changes, dy, with infinitesimal change, dx, in the independent variable, x. Such is expressed mathematically as $dy/dx = \lim_{\Delta x \to 0} (\Delta y/\Delta x)$.

Addressing problem #7 through the RMT model explicitly reveals the systemic nature of mathematics concepts, from basic math through geometry, trigonometry, algebra and leading to calculus. There is a chain or network of relationships formed through the conceptual components of quantities, points, line segments, angles, perpendiculars, right triangles, ratios, proportions, variables, rate, slope, derivative. ■

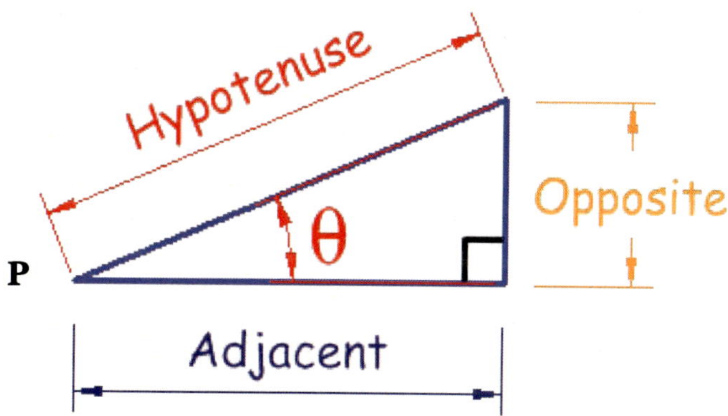

Chapter Six

Applying Cognitive Functions and Mathematical-Physics Concepts to Electrical Circuitry

In This Chapter

- Contributed by RMT Laboratory Learners
  Carolyn Whitehead-Hooks and Craig Washington

Enhancing Abstract Representational-Relational Thinking Through FIE Instrument: Representational Stencil Design

Wisdom is the principal thing; Therefore get wisdom.
And in all your getting, get understanding.
~Proverbs 4:7

Applying the RMT cognitive function functions that create abstract representational-relational thinking is essential for mathematical-physics conceptual understanding and complex problem solving. Demonstration of strategic competence through such problem solving requires complex levels of representational internalized behavior. The FIE instrument Representational Stencil

The Set of Stencils to Use

Design (RSD) provides a rich battery of general psychological tools that promote practice in the spontaneous integration and dynamic use of these essential cognitive functions. The cognitive tasks in RSD engage the learner in mentally constructing a design that is identical to that in a colored standard. Each design comprises a complex figure constructed of shapes and colors, which are the composite of an interplay of the segregation of the lines and colors of different stencils (Feuerstein and Hoffman, 1995, p. 2).

Design One

Analysis:

The design consists of a blue center cross contained within a yellow border.

Starting at the center of the design, I focus on color and shape. I find a blue cross. Looking at my set of stencils, I determine that my foundational color must contain the color blue. Searching my stencils, I select stencil #5.

Next, I look for a stencil that, when placed over stencil #5, will contribute to the selected design. I choose stencil #14 because it has my required shape of a cross and color yellow.

This same system of analyzing, decomposing, looking for relevant cues, and superimposing can be used to complete other designs—for example, see Design Two below, which required more rigor in the use of the cognitive functions.

Design Two

Analysis:

This design features a white center cross surrounded by a larger green cross bor-dered by a larger black cross enclosed within a green octagonal shape with red corners contained within a black outer border.

Applying the RMT cognitive function of analyzing, I started by spending some time analyzing the final design and its composition. In order to determine the sequence of stencils needed to create the design, I decomposed it into its parts. I looked for relevant cues such as color, shape, orientation, and location. As I look for these attributes, I then analyze the set of stencils shown below to determine which one will contribute to the current level of the design. I utilize the cognitive function of superimposing to develop a picture in my mind of what the design will look like after adding each stencil.

Analysis and Integration:

1. Foundational white center. Stencil 15.
2. Green with center cross. Stencil 12.
3. Red at corners with octagon. Stencil 9.
4. Black borders, diagonal cross. Stencil 3.

This requires the combined use of the cognitive functions labeling, visualizing, conserving constancy, comparing, forming relationships, providing logical evidence, and integrating.

Inductance:

Applying the RMT cognitive function

Using a Diagram to Unpack the Structure of the Problem

 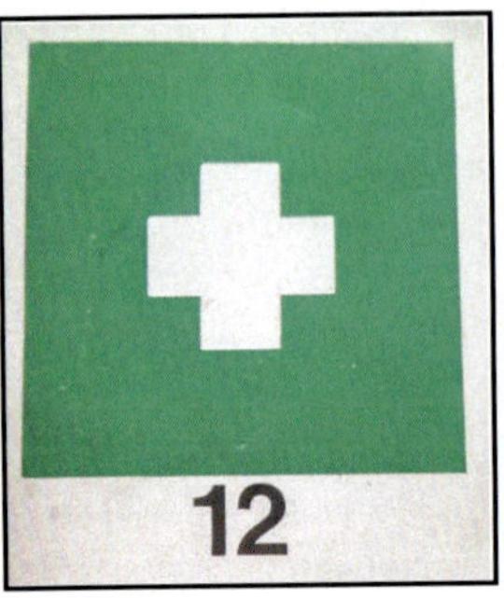  

analyzing, I began by spending some time investigating and decomposing the concept of Inductance, which is also referred to as Electromagnetic Induction. In order to understand Inductance, I had to first decompose it into its physical characteristics/attributes. The physical interaction of a coil of wire with a magnetic core having a current applied to it generates a counter electromagnetic force or emf. This counter emf then opposes the current. The relationship between this counter emf and the current represents the main concept of inductance. Understanding this relationship is key to understanding the mathematical relationships between the emf and the current.

The emf is the induced voltage produced when the magnetic flux in a coil is changed. This magnetic flux is the amount of magnetic field lines that pass through an area. The inductance created by the coil can be modified by changing the surface area of the coil, the number of turns in the coil and changing the current.

This mathematical relationship is seen in Faraday's Law, which relates the rate of change of the magnetic flux through a coil to the magnitude of the emf induced in the coil.

$$EMF = N\frac{d\phi}{dt}$$

where: $d\phi$ = the change in the magnetic flux

N = the number of turns of the coil

Lenz's Law also defines a mathematical relationship that relates the direction that the current will flow. It states that the direction will oppose the change in flux that produced it. This means that the magnetic field produced by an induced current will be in the opposite direction to the change in the original field. This is why the negative sign is added to the equation.

$$EMF = -N\frac{d\phi}{dt}$$

where: $d\phi$ = the change in the magnetic flux

   N = the number of turns of the coil

Activating this newly acquired understanding of Inductance, I was able to analyze and solve the circuit below using the same RMT cognitive functions.

Problem #1

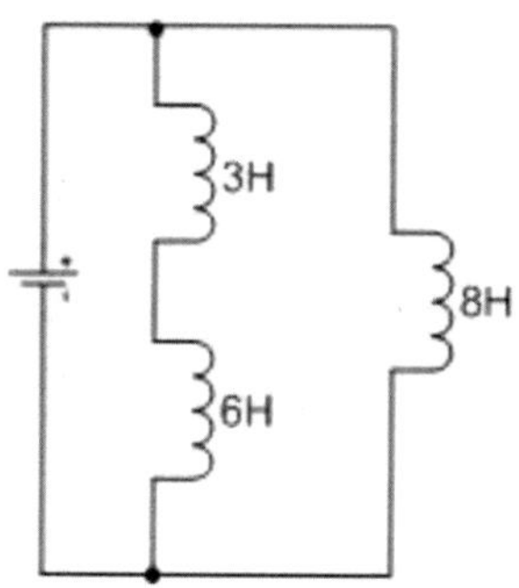

What is the total inductance of the circuit below?

We begin by using our cognitive function of analyzing. This step allows us to activate any prior knowledge of the subject, which may lead us to the next cognitive function of defining the problem. We proceed to define the problem by identifying the known and the unknown.

<u>Known</u>
Series – parallel circuit
L_1 = 3H (series)
L_2 = 6H (series)
L_3 = 8H (parallel)

<u>Unknown</u>
Ltotal = ?

where:
 L = Inductance
 H = Henry (unit of measure for inductance)

In order to find the logical path from the known to the unknown we analyze what we know already from the known and then try to construct a logical path to the unknown.

Activating prior knowledge about inductors in a circuit, we first need to determine how they are connected.

Looking at the series section of the circuit, we determine that L_1 and L_2 are connected in series. When inductors are connected in series, the total inductance is the sum of the individual inductor quantities.

Inductance for inductors connected in Series:

$$Lseries = L_1 + L_2 + L_3 + \dots L_n$$

So...

$$L(1 \text{ and } 2) = L_1 + L_2$$
$$= 3H + 6H$$
$$= 9H$$

Looking at our circuit, we now have L(1 and 2) and L_3 connected in parallel.

When inductors are connected in parallel, the total inductance is less than any one of the parallel inductors quantities. We calculate Total parallel inductance similar to the way that we calculate parallel resistance. We find the reciprocal of the sum of the reciprocal individual inductor quantities.

Inductance for inductors connected in Parallel:

$$Lparallel = 1\ (1/L_1 + 1/L_2 + 1/L_3 + \ldots + 1/L_n)$$

Or (if 2 values)

$$Lparallel = (L(1\ and\ 2) * L_3)/(L(1\ and\ 2) * L_3)$$

So…

$$Lparallel = (9H * 8H) / (9H + 8H)$$
$$= 72H2/17H$$
$$= 4.24H$$

(answer) Ltotal = 4.24H

Problem 2

In the circuit below, what is the calculated power for R_2?

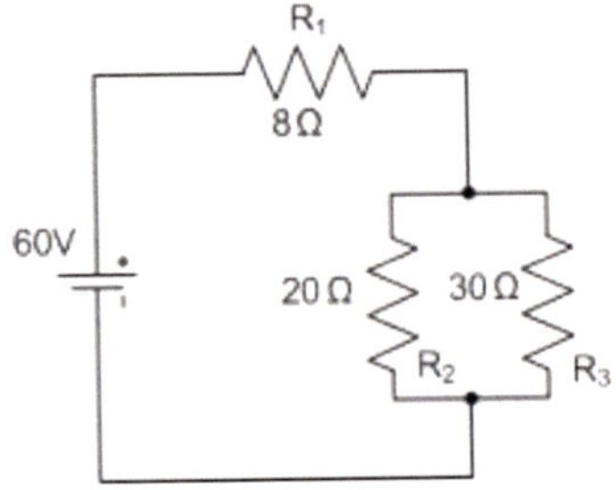

Systematically searching for clear and complete information, it is fair to say that this circuit is a combination of a series and a parallel circuit. We can ask ourselves, how did we come to that conclusion? By analyzing the field of data, activating prior knowledge, forming relationships, and providing logical evidence, we identify the critical attributes of a series circuit and a parallel circuit.

Series Circuit

a. components are connected end-to-end
b. there is only one pathway for current to flow in a branch

Parallel Circuit

a. components can be connected
b. a circuit that contains more than one pathway for current to flow

We observe that Resistor 1 is connected end-to-end and there is only 1 pathway for current to flow. Conclusion: Resistor 1 is in series. However, Resistor 2 and Resistor 3 are not connected end-to-end but are branched and allow current to flow in two directions. Conclusion: Resistor 2 and Resistor 3 are in parallel with respect to each other.

Note: (I say with respect to each other because the branch that contains Resistor 2 and Resistor 3 are connected end-to-end with Resistor 1. Thus, they are in series with Resistor 1). This concludes what we observed about the circuit in the problem. Now to use the cognitive function defining the problem.

We have previously used several of the cognitive functions such as comparing, focusing, labeling, superimposing, etc., but in problem solving of this nature, we must use the cognitive function defining the problem. We will:

a. Analyze—focus on a situation/problem, breaking it down into components, searching for discrepancies, differences or variances etc. and write them down!

b. Compare—distinguish what information is "known" or "given" in the problem from what is "unknown" or what the problem is asking for. Known information can be formulas, diagrams, critical attributes, etc. Write them down!

c. Construct the most "logical path" to move from what is "known" to the "unknown."

<u>Known</u>   <u>Unknown</u>
A series – parallel circuit $P_{R3} =$
$V_T = 60V$
$R_1 = 8\Omega$
$R_2 = 20\Omega$
$R_3 = 30\Omega$

<u>Constructing a Logical Pathway from the Known to the Unknown</u>

We know that $P_{R3} = I_{R3} \times V_{R3}$
We also know that $V = I \times R$ and that I

$= V/R$, but we need I_{R3}, which is in the parallel segment of the circuit. We can simplify this circuit by resolving R_2 and R_3, which are in parallel, into one resistor value. The circuit can then be treated as a series circuit with just resistors R_1 and $R_{2+3}.$*

***Note:** The notation "R_{2+3}" represents the combined value of R_2 and R_3 in parallel.

From the critical attributes of resistors in a parallel circuit, we know how to calculate the R_{2+3} resistor value for $R_2 = 20\Omega$ and $R_3 = 30\Omega$

$$R_{2+3} = 1/(1/R_2 + 1/R_3) = 1/((1/20\Omega) + (1/30\Omega)) = 12\ \Omega$$

Now we can calculate the total resistance (R_T) for our improvised series circuit. From our critical attributes of a series circuit we know the total resistance (R_T) of the circuit is the sum of the individual resistors:

$$R_T = R_1 + R_2 + R_3,$$

We also know $R_1 = 8\ \Omega$ and $R_{2+3} = 12\ \Omega$ Therefore $R_T = 8\Omega + 12\Omega = 20\Omega$

Now we can calculate the total current (I_T) for our improvised series circuit. We know voltage $(V_T) = 60$ V and we know total resistance (R_T) of our improvised circuit.

$$I_T = V_T / R_T \qquad 60\ V/20\ \Omega = 3\ A$$

From the critical attributes of a series circuit, we know that the sum of the voltage drops across all resistors conserves constancy with the total voltage. We also know from the critical attributes of a series circuit that the total current conserves constancy throughout all resistors.

a. $V_T = V_1 + V_2 + V_3$
b. I_T is constant or stays the same throughout the circuit

Logical Plan

If we can find the voltage drop across resistor 1 (R_1), then we can subtract it from the total voltage to find the voltage at R_{2+3}

a. $V_1 = I_T \times R_1$, $V_1 = (I_T = 3A) \times (R_1 = 8\,\Omega) = 24$ V
b. $V_{2+3} = V_T - V_1 = 60v - 24v = 36v$

From the critical attributes of a parallel circuit, we know that voltage conserves constancy throughout the components or it stays the same in the components that exist in a parallel relationship. Now let's go back and look at the information we have attained and re-examine the parallel portion of the circuit.

<u>We have:</u>
$R_2 = 20\,\Omega$
$R_3 = 30\,\Omega$
$V_{2+3} = 36$V and is constant or the same through both Resistor 1 and Resistor 2. We know from Ohm's Law that:

Current $(I) = V/R$. Therefore .... $I_2 = V_{2+3} / R_2$ and .... $I_3 = V_{2+3} / R_3$

a. $I_2 = V_{2+3} / R_2$ or $36V/20\Omega = 1.8A$
b. $I_3 = V_{2+3} / R_3$ or $36V/30\Omega = 1.2A$

Now we have all of the data we need to calculate power in the three resistors. We know that power $(P) = $ current $(I) \times$ voltage (V)

$P_3 = I_3 \times V_{2+3}$ $P_3 = (1.2A)(36V) = 43.2$ Watts
$P_2 = I_2 \times V_{2+3}$ $P_2 = (1.8A)(36V) = 64.8$ Watts
$*P = I_T \times V_1$ $P_1 = (3.0A)(24V) = 72.0$ Watts

Remember: Resistor 1 is in the series portion of the circuit and current (I) conserves constancy, $I_T = I_1$. We have solved this problem while providing all of our mathematical-scientific logical evidence. ■

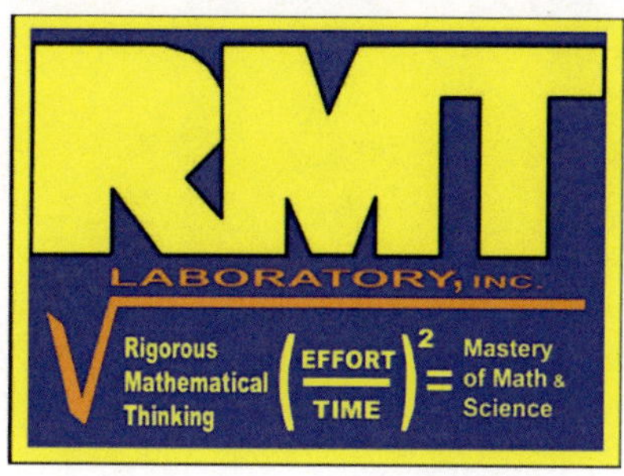

The RMT Mission

RMT Laboratory is an analytical-critical thinking laboratory that engages learners in rigorous mathematical thinking to learn mathematics the way mathematicians do mathematics. Our focus is developing mathematical proficiency in big mathematical ideas, from basic math through calculus. Our definition and measure of mathematical proficiency in each big idea are based on the five strands delineated by the National Research Council (NRC, 2001). These five strands for mathematical proficiency are deep conceptual understanding, strategic competence, adaptive reasoning, procedural fluency, and productive disposition.

The goal of this mission is to expand the STEM (science, technology, engineering, and mathematics) pipeline for all committed learners so they can:

- be successful through high school and college;
- be analytical-critical thinkers;
- be complex real-world problem solvers; and,
- compete in this global technological economy.

About the Authors

James T. Kinard Sr. earned a PhD in theoretical and experimental electroanalytical chemistry and an MS in theoretical physical chemistry from Howard University. He also earned a BS in chemistry and mathematics from Benedict College. He has been a professor of chemistry and principal investigator of research sponsored by the National Aeronautics and Space Administration (NASA), U.S. Environmental Protection Agency (EPA), National Institutes of Health, U.S. Bureau of Mines, and Agency for International Development. He was also an analytical chemist at Savanah River Laboratory (U.S. Atomic Energy Commission) and Lawrence Livermore National Laboratory (U.S. Department of Energy). He is a certified trainer of the Feuerstein cognitive development program, Instrumental Enrichment, and has lectured at international cognitive enrichment workshops in the United States, Canada, the United Kingdom, France, The Netherlands, and India. Dr. Kinard has published numerous research articles in refereed scientific journals, articles on Rigorous Mathematical Thinking in mathematics education and psychology journals and is first author of *Rigorous Mathematical Thinking: Conceptual Formation in the Mathematics Classroom* (Cambridge University Press, 2008).

Gwendolyn D. Gibson Kinard earned a PhD in public policy analysis, with a focus on education administration, from the University of Illinois at Chicago. She earned an MA in early childhood special education from Northeastern Illinois University and a BA in chemistry from Concordia College. She taught mathematics and science at the high school level and was an administrator for Chicago Public Schools' National Science Foundation funded Chicago Systemic Initiative, a precursor of STEM (science, technology, engineering, and mathematics) initiatives and the Common Core Standards. She is a certified trainer of the Feuerstein cognitive development program, Instrumental Enrichment.

References

Books

Baxter, J. M., Moir, A. & Rorie, J. eds. (1910). *Exhortations.*
Dundee: John Leng and Co., Ltc., pp. 5, 16, and 18.

Campbell, T. M. (1936). *The Movable School Goes to the Negro Farmer.* Tuskegee, AL: Tuskegee Institute Press

Campbell, L., & Garnett, W. (1882). *The Life of James Clerk Maxwell: with a Selection from His Correspondence and Occasional Writings and a Sketch of His Contributions to Science.* London: Macmillan and Co. 1882. pp.513-556.

Carver, George Washington (1905). "How to Build Up Worn Out Soils," *Tuskegee Experiment Station, Bulletin Six* (Tuskegee, 1905, p. 4.)

Falik, L. H. (In press). *Changing Destinies: The Extraordinary Life of Reuven Feuerstein*

Farady, Michael. *The World Book Encyclopedia* (1999). Chicago:World Book Inc. p. 63.

Federer, W. J. (1994). *America's God and Country Encyclopedia of Quotations.* Coppell, TX: FAME Publishing, p. 96.

Felix, J. (1971). *The Tuskegee Institute Movable School, 1906–1923.* Agricultural History, 45, pp. 201-209.

Feuerstein, R., Feuerstein, R.S., & Falik, L. (2010). *Beyond Smarter: Mediated Learning and the Brain's Capacity for Change.* New York, N.Y.: Teachers College Press

Feuerstein, R., Feuerstein, S., Falik, L., & Rand, Y. (1979; 2002). *Dynamic Assessments of Cognitive Modifiability.* Jerusalem, Israel: ICELP Press

Feuerstein, R., Feuerstein, R. S., Falik, L. H., & Rand, Y. (2006). *The Feuerstein Instrumental Enrichment Program.* Jerusalem, Israel: ICELP Publications.

Feuerstein, R. Rand, Y., Hoffman, M.B., & Miller, R. (1980; 2004). *Instrumental Enrichment: An Intervention Program for Cognitive Modifiability.* Baltimore, MD: University Park Press.

Feuerstein, R. and Hoffman, M. (1995). *Teacher's Guide to Categorization.* Palatine, IL: IRI SkyLight.

Feuerstein, R., Rand, Y., Hoffman, M., & Miller, R. (1980). *Instrumental Enrichment.* Baltimore, MD: University Park Press. pp. 116, 273

Feuerstein, R. and Hoffman, M. (1995). *Teacher's Guide to Numerical Progressions,* IRI SkyLight.

Feuerstein, R. and Hoffman, M. (1995). *Teacher's Guide to Representational Stencil Design,* IRI SkyLight.

Feuerstein, R., Rand, Y., & Rynders, J. E. (1988). *Don't Accept Me as I am: Helping "Retarded" People to Excel.* New York, NY: Plenum Press.

Feuerstein, R., Klein. P.S. and Tannenbaum, A. J., eds. (1991). *Mediated Learning Experience (MLE): Theoretical, Psychosocial and Learning Implications.* Freund Publishing House Ltd.

Gibran, K. (1923). *The Prophet*. New York, NY: Alfred A. Knopf.

Kinard, J. and Kozulin, A. (2008). *Rigorous Mathematical Thinking: Conceptual Formation in the Mathematics Classroom*. Cambridge, UK: Cambridge University Press.

Kozulin, A. (1998). *Psychological Tools: a Sociocultural Approach to Education*. Cambridge, MA: Harvard University Press.

Kremer, Gary R., ed. (1987). *George Washington Carver in His Own Words*. Columbia, Missouri: University of Missouri.

McMurry, L. (1981). *George Washington Carver: Scientist & Symbol*. Oxford: Oxford University Press, pp. 9-10.

Pink, A.W. (1930). *The Attributes of God*. Pensacola, Florida: First Chapel Library.

Shetterly, M.L. (2016). *Hidden Figures: The American Dream and the Untold Story of the Black Women Mathematicians Who Helped Win the Space Race*. New York, NY: William Morrow.

Skuy, M. (1996). *Mediated Learning In and Out of the Classroom*. Arlington Heights, IL: IRI/SkyLight Training and Publishing, Inc.

Thurman, H. (1996). *Jesus and the Disinherited*. Boston, MA: Beacon Press.

Tyler, R.W. (1949). *Basic Principles of Curriculum and Instruction*. Chicago, IL: The University of Chicago Press.

Warren, Wini (1999). *Black Women Scientists in the United States*. Bloomington, IN. [u.a.]: Indiana University Press. pp. 148–150.

Articles

Dao, C. (2008). "Man of Science, Man of God: George Washington Carver." *Act & Facts*. 37 (12): 8.

Eichman, P. (1993). Michael Faraday. Perspectives on Science and Christian Faith 43 (June 1993): 92-95.

Feuerstein, R., and Falik, L. H. (2010). "Learning to think, thinking to learn: A comprehensive analysis of three approaches to instruction." *Journal of Cognitive Education and Psychology*, 9(1), 4-20. doi: 10.1891/1945-8959.9.1.4

Feuerstein, R. (1990). "The theory of structural modifiability." In B. Presseisen (Ed.), Learning and thinking styles: Classroom interaction. Washington, DC: National Education Association.

Hellman, H. (1998). "Two views of the universe—Galileo vs. the Pope." *Washington Post*, September 08, 1998, page H01, The Washington Post Company.

Jones, A. W. (1975). "The Role of Tuskegee Institute in the Education of Black Farmers." *Journal of Negro History* 60, 252-67.

Kinard, J. T. (2001). "Cognitive, affective and academic changes in specially challenged inner-city youth." In A. Kozulin, R. Feuerstein, and R.S Feuerstein (Eds.), Mediated learning experience in teaching and counseling (pp. 55 – 64). Jerusalem: ICELP Press.

REFERENCES

Kinard, J.T. (2006). "Creating Rigorous Mathematical Thinking: A Dynamic that Drives Mathematics and Science Conceptual Development." *Transylvanian Journal of Psychology*, [Special Issue] No. 2, pp 251 – 266.

Kinard, J.T. and Falik, L. (1999). "The fourth r: creating rigorous thinking through mediated learning experience and Feuerstein's instrumental enrichment \ program," *Life Span and Disability*, 2(2): 185-204.

Kinard, J.T. and Kozulin, A. (2005). "Rigorous mathematical thinking: mediated learning and psychological tools." *Focus on Learning Problems in Mathematics*, Vol. 27, #3.

Lamont, Ann (1990). Michael Faraday—"God's Power and Electric Power." *Creation* 12 (4): 22-24.

National Council of Teachers of Mathematics. (2000). "Principles and standards for school mathematics." Reston, VA: NCTM *Newport News*.

Obituary: Dorothy Vaughan, "Newport News,"November 2008.

Rand, Y. (1991). "Deficient cognitive functions and non-cognitive determinants—An integrating model: assessment and intervention." In R. Feuerstein, P. Klein, and A. Tannenbaum (Eds). Mediated learning experience (MLE.): Theoretical, psychological and learning implications. p. 80 - 81). Jerusalem: ICELP Press

Toom, A. (1993). "A Russian teacher in America." *American Education*, 17(3): 9-13, 20-25.

Internet Citations

American Chemical Society National Historic Chemical Landmarks. George Washington Carver: Chemist, retrieved from: http://www.acs.org/content/acs/en/ education/whatischemistry/landmarks/carver.html.

Angie Turner King (n.d.). Retrieved from Wikipedia on https://en.wikipedia.org/wiki/ AngieTurnerKing

Bagley, M. (2013). George Washington Carver: Biography, Inventions & Quotes. Retrieved from LiveScience: http://www.livescience.com/41780-george-washi

Biography.com Editors. (n.d.). Dorothy Johnson Vaughan biography.com. Retrieved June 9, 2017 from The Biography.com website: https://www.biography.com/ people/dorothy-johnson- vaughan-111416

Blair, E. (2016). "Hidden Figures" no more: Meet the black women who helped send America to space. Retrieved June 9, 2017 from npr web site: http://www. npr.org/2016/12/16/505569187/hidden-figures-no-more-meet-the-black-women-who-helped-send-america-to-space

Blaise Pascal, (n.d.). Retrieved June 9, 2017 from Wikipedia website: https://en.wikipedia. org/wiki/Blaise_Pascal

Carver, A Brief Sketch of My Life. George Washington Carver, 1897 or thereabouts, George Washington Carver Papers, Tuskegee Institute Archives, reel 1. https:// www.austint exas.gov/sites/default/files/files/Parks/Carver_Museum/Carver_ Bio_a

Coble, D. (2017). Hidden Figures mathematician raised Methodist. White Sulphur Springs, W.VA (UMNS). Retrieved from The People of the United Methodist Church http://www.umc.org/news-and-media/hidden-figures-mathematician-raised-methodist

Collins, S.N. (2016). Unsung: William Claytor. Retrieved June 15, 2017 from Undark website: https://undark.org/article/unsung-william-waldron-schieffelin-claytor/

Colorado Department of Education: STEM standards and instruction https://www.cde.state.co.us/stem/standards

Devlin, K. (2017). All the mathematical methods I learned in my university math degree became obsolete in my lifetime. The Huffington Post, 01/01/17 Retrieved on February 3, 2017 from http://www.huffingtonpost.com/entry/all-the-mathematical-methods-i-learned-in-my-university_us_58693ef9e4b014e7c72ee248

DeSilver, D. (2017). U.S. students' academic achievement still lags that of their peers in many other countries. Pew Research Center. http://www.pewresearch.org/fact-tank/2017/02/15/u-s-students-internationally-math-science/

Dorothy Vaughan (n.d.). Retrieved June 9, 2017 from Wikipedia website: https://en.wikipedia.org/wiki/Dorothy_Vaughan

Eichman, P. (1991). The Christian Character of Michael Faraday as Revealed in His Personal Life and Recorded Sermons. Retrieved from: http://www.asa3.org/ASA/PSCF/1991/PSCF6-91Eichman.html

Eichman, P. (1993). Michael Faraday. *Perspectives on Science and Christian Faith* 43 (June 1993): 92-95.

Famous Scientists Who Believed in God, (n.d.) Retrieved June 6, 2017 from Evidence for God http://www.godandscience.org/apologetics/sciencefaith.html

Faraday's Line of Force and Maxwell's Theory of the Electromagnetic Field. (n.d.). Retrieved June 17, 2017 from In Compliance website: http://incompliancemag.com/article/faradays-lines-of-force-and-maxwells-theory-of-the-electromagnetic-field/

"Former NASA Langley Mathematician to be Awarded Presidential Medal of Freedom" (Nov. 17, 2015). Retrieved June 14, 2017 from NASA History website: https://www.nasa.gov/feature/katherine-johnson-the-girl-who-loved-to-count

Galileo, Astronomer, Scientist (n.d.). Retrieved June 8, 2017 from Galileo Biography.com website: https://www.biography.com/people/galileo-9305220

Galileo in Rome for Inquisition (n.d.). Retrieved June 8, 2017 from This Day in History website: http://www.historycom/this-day-in-history/galileo-in-rome-for-inquisition

REFERENCES

George Washington Carver, n.d. Retrieved June 16, 2017 from Wikipedia website: https://en.wikipedia.org/wiki/George_Washington_Carver

Ghose, T. (2016). "The human brain's memory could store the entire Internet." Retrieved from Live Science http://www.livescience.com/53751-brain-could-store-internet.html

Ginsberg, R. (2005). "The Transformer." Retrieved June 17, 2017 from aish.com website: http://www.aish.com/jw/id/48914587.html

Internet of the Mind (2007 - 2017). Retrieved from: http://www.internet-of-the-mind.com/neural_networks.html

"Katherine Johnson: A Lifetime of STEM" (2013). Retrieved June 14, 2017 from NASA Langley website: https://www.nasa.gov/audience/foreducators/a-lifetime-of-stem.html

Katherine Johnson (n.d.) retrieved June 15, 2017 from Wikipedia website: https://en.wikipedia.org/wiki/Katherine_Johnson

Kerr, J. (2016). "Internationally, U.S. students are falling." Associated Press. Retrieved from https://www.usnews.com/news/politics/articles/2016-12-06/math-a-concern-for-us-teens-science-reading-flat-on-test

Khron, A. (2017). "Anna Krohn: Hidden Partnerships—Women, Science, and Faith." *The Catholic Weekly.* April 12, 2017. Retrieved from: https://www.catholicweekly.com.au/anna-krohn-hidden-figures-partnerships-women-science-faith/

Kiefer, J.E. (n.d.). "Blaise Pascal, scientist, religious writer." Retrieved June 8, 2017 from Biographical Sketches website: http://justus.anglican.org/resources/bio/233.html

Kilpatrick, J., Swafford, J., & Findell, B. (2001). "Adding it up: Helping children learn mathematics." Washington, DC: National Academy of Sciences – National Research Council. Retrieved from http://www.nap.edu/catalog/9822.html

King, A. (1955). "An analysis of early algebra textbooks used in the American secondary schools before 1900." University of Pittsburgh. Retrieved 21 April 2017 from blackpast@blackpast.org

Kozulin, A. (2001). "Mediated learning and cultural diversity." International Center for the Enhancement of Learning Potential. http://www.umanitoba.ca/unevoc/conference/papers/kozulin.pdf

Loff, S. (November 22, 2016). Katherine Johnson Biography. NASA. Retrieved February 1, 2017 from https://www.nasa.gov/content/katherine-johnson-biography

Mary Jackson (engineer), (n.d.). Retrieved June 13, 2017 from Wikipedia website: https://en.wikipedia.org/wiki/Mary_Jackson_(engineer)

Mary Winston Jackson Obituary (2005). Retrieved June 13, 2017 from Daily Press Obituaries website: http://www.legacy.com/obituaries/dailypress/obituary.aspx?pid=3163015

Michael Faraday (2014). Famous Scientists. Retrieved June 17, 2017 from famousscientists. org. website: www.famousscientists.org/michael-faraday/

Mirk, S. (2015). "African American women worked as some of NASA's first computers." Retrieved June 9, 2017 from Bitchmedia website: https://www.bitchmedia.org/article/african-american-women-worked-some-nasas-first-computers

National Council for Teachers of Mathematics (2000). Retrieved March, 2017 from http://www.nctm.org/Standards-and-Positions/Principles-and-Standards/

Obituary: Dorothy Vaughan, "Newport News." November 2008. O'Conner, J.J. and Robertson, (2016). Katherine Coleman Goble Johnson. Retrieved June 14, 2017 from MacTutor History of Mathematics website: http://www.groups.dcs.st-and. ac.uk/history/Biographies/Johnson_Katherine.html

Overman, C. (2010). George Washington Carver. Retrieved June 16, 2017 from Transforming Teachers website: https://transformingteachers.org/en/articles/biblical-integration/stories-for-teachers/294-george-washington-carver

Parsons, J. (n.d.). The Hebrew name for God http://www.hebrew4christians.com/ Names_of_G-d/YHVH/yhvh html

Pascal's triangle (n.d.). Retrieved June 9, 2017 from Wikipedia on https://en.wikipedia. org/wiki/Pascal%27s_triangle

Reuven Feuerstein, (n.d.). Retrieved from Wikipedia website: https://en.wikipedia.org/ wiki/Reuven_Feuerstein

Selected Exhortations Delivered to Various Churches of Christ by the Late Michael Faraday Retrieved on July 29, 2017 from http://www.asa3.org/ASA/ PSCF/1991/PSCF6-91Eichman.html

[A]Shetterly, M.L. (n.d.). Dorothy Vaughn biography. Retrieved June 9, 2017 from NASA website: https://www.nasa.gov/content/dorothy-vaughan-biography

[B]Shetterly, M. L. (n.d.). Mary Jackson biography. Retrieved June 9, 2017 from NASA website: https://www.nasa.gov/content/mary-jackson-biography

Shetterly, M. L. (December 1, 2016). "From Hidden to Modern Figures—Katherine Johnson Biography." NASA. Retrieved March 1, 2017 from https://www.nasa. gov/

Smith, Yvette (November 24, 2015). "Katherine Johnson: The Girl Who Loved to Count." NASA. Retrieved February 12, 2016 from https://www.nasa.gov/feature/ katherine-johnson-the-girl-who-loved-to-count.

Sonia (March 15, 2017). "Katherine G. Johnson: NASA Mathematician and Dedicated Presbyterian." Presbyterian Historical Society. https://www.history.pcusa.org/ blog/2017/03/katherine-g-johnson-nasa-mathematician-and-dedicated-presbyterian

The Pittsburgh Courier. (1955, February 12). "Ph.D. Degree to Mrs. King." Retrieved 18 February 2017.

William Claytor, a mathematical genius (n.d.). Retrieved June 15, 2017 from African American Registry we site http://www.aaregistry.org/historic_events/view/

william-claytor-mathematical-genius.

Williams, L.P. n.d. "Michael Faraday—British physicist and chemist." https://www.britannica.com/biography/Michael-Faraday

Zabawa, R. (2008). "Tuskegee Institute movable school." Encyclopedia of Alabama (website): http://www.encyclopediaofalabama.org/article/h-1870

Quotations

Adams, H.B. (n.d.). BrainyQuote.com. Retrieved June 6, 2017, from BrainyQuote.com website: https://www.brainyquote.com/quotes/quotes/h/henryadams100161.html

Edison, Thomas http://www.goodreads.com/quotes/551-five-percent-of-the-people-think-ten-percent-of-the

Ford, Henry (2016). Xplore Inc., Retrieved December 6, 2016from BrainyQuote.com website: https://www.brainyquote.com/quotes/quotes/h/henryford122851.html.

King, Jr., Martin Luther (2016). BrainyQuote.com. Xplore Inc. 6 December 6, 2016. https://www.brainyquote.com/quotes/quotes/m/martinluth136311.html

Wilde, O. (n.d.). BrainyQuote.com. Retrieved June 6, 2017, https://www.brainyquote.com/quotes/quotes/o/oscarwilde161644.html

91260810R00111

Made in the USA
Lexington, KY
19 June 2018